THE GRAPHENE BRIDGE

KIREET KOKALA &
SUPANTHA BANERJEE

Disclaimer

The information contained in this book is for general informational purposes only, with fictitious sections presented in *italics*. Advanced technologies including AI tools have been carefully used to augment research, enhance imagination, and streamline the creative writing process. Except for figures and tables, the artwork was enhanced with AI. All facts and resources have been thoroughly reviewed, cited, refined, and edited to ensure accuracy, originality, and quality. While every effort has been made to ensure the reliability of the information presented, the authors make no representations or warranties of any kind, express or implied, about the completeness, accuracy, suitability, or availability of the content contained in this book. The authors disclaim any responsibility for errors, omissions, or damages arising from the use of or reliance on this information. Readers are encouraged to consult qualified professionals for specific questions related to the topics discussed in this book.

ISBN: 979-8-9985483-0-7 (paperback)

Advance Praise for The Graphene Bridge

“A thriller-like multinational story of scientific discovery and development, spanning continents, leveraging exotic locales, peppered with intriguing characters, is set against a backdrop of high stakes and extreme urgency.”

“The Graphene Bridge center-stages the eponymously named material. It surrounds graphene with rich context, using Artificial Intelligence (AI), Quantum Computing, Atomic Batteries and Cybersecurity. Using these four contextual components to frame an engaging story, the authors have stitched a tapestry of use cases.”

— **Mr. Ashwin Rangan**, Chief Executive Officer at DoubleCheck Solutions.

"The fictional story is a clever way of keeping the interest of a popular science reader. It breaks up the technical narrative and hopefully maintains the reader’s interest in getting to the end of the story (and the end of the book)."

"The target audience for this book is a popular science readership. Someone who has little formal training or education in STEM disciplines or hasn’t used STEM skills in their everyday jobs for a long time but nevertheless has an interest in new technology developments."

— **Mr. Mark Settle**, Former CIO and Author of *Truth from the Valley* and *Truth from the Trenches*.

"Forward-thinking and innovative, The Graphene Bridge offers a provocative look at the untapped potential and applications of a groundbreaking material to our everyday life and the future of human society. It intricately weaves graphene through time and themes and truly sparks imagination and wonderment. Whether you are a tech enthusiast or simply intrigued, this book will leave your curiosity piqued and wanting more."

— **Mrs. Kathyrn N. Guzner**, Attorney and Founder of Guzner Law, P.C.

"The core strength of this book is the eye-opening perspectives on the endless possibilities graphene materials can deliver to make our society more productive and sustainable. Most individuals (including myself) have no idea this material even existed. This book clearly defines this."

"I like how the educational component of graphene technology is weaved with a suspenseful narrative on how it can be incorporated. This provides greater insight into how this new material can make a tremendous impact on technology and infrastructures."

— **Mr. Kevin Kokinda**, Resumedics Founder and Author of *Finding Your Career Mojo: Muddling thru the Resume' BS.*

"The authors' dedication to exploring complex ideas and presenting them in an accessible format is evident throughout the work."

"This book demonstrates a commendable effort in addressing a significant subject matter."

— **Dr. Abhimanyu Gupta**, Instructor, Department of Operations & IT Management at Richard A. Chaifetz School of Business, Saint Louis University.

Dedications

To my wonderful parents, Rajani and Satya, your unwavering encouragement fuels my aspirations. To my mother, Rajani, I'm humbled to observe you have remained my toughest critic and strongest supporter. To my loving wife, Ekta, your conversations help anchor me like a compass to true north. And to my amazing children, Isha and Veda, you make my heart tick with joy. Lastly, thanks to my college professors, martial arts instructors, and venerable gurus who buttressed the value of lifelong curiosity.

My heartfelt gratitude is due to my fellow author Supantha, for embarking on a profound and creative journey with me. Your collaboration, critical thinking, insights, and counsel have enriched every bit of this incredible endeavor.

As I share these sentiments, a piercing quote comes to mind:

'When ambitions are high, half-empty stomachs run twice full.'

Dear readers, may we stay hungry and creative in the face of discovery.

Kireet

"The most profound journeys are never solitary—they are woven from the threads of love, sacrifice, and shared dreams."

To my parents, who first showed me the power of perseverance and belief.

To my wife, Puspita, whose love and partnership have been my constant source of strength and inspiration.

To Reina and Ronav, my greatest teachers and most cherished sources of joy — may this book spark your own boundless curiosity.

And to my co-author, Kireet, whose brilliant mind and unwavering commitment transformed our shared vision into reality.

Supantha

Contents

Preface

Embark on the pages of The Graphene Bridge, a journey exploring the transformative potential of graphene. This book probes the intersection of graphene with Artificial Intelligence (AI), Quantum Computing, and Atomic Batteries, examining its role in advancing these critical technologies. It welcomes technology fans of all backgrounds by breaking down complex scientific ideas into understandable concepts. Whether you're a curious layperson or a professional seeking to broaden your understanding of cutting-edge developments, you'll find this exploration informative and thought-provoking.

Imagine a future where AI seamlessly integrates with the quantum realm, and energy sources are both abundant and sustainable. Graphene, with its extraordinary properties, exceptional strength, unparalleled conductivity, and remarkable flexibility, has the potential to unlock these and many other technological marvels.

The non-fiction book sections are merged with a *fictitious* conceptualization (*presented in italics*) and offered through a hybrid lens. Inspiration for the work largely comes from two seminal scientific papers from the 1980s: the late Fred Brooks' "No Silver Bullet" and Ken Thompson's "Reflections on Trusting Trust."

By the end of this journey, you'll gain a deeper understanding of graphene's potential to revolutionize the global landscape, from the microscopic to the macroscopic, and its profound impact on the future of humanity. You are invited to consider the possibilities, ponder the challenges, and ultimately, become part of the conversation about the future of graphene and its role in shaping our world.

Foreword

At first, when I was requested to write the Foreword for this book, I was somewhat taken aback. After all, my work history is dealing with and managing information technologies, admittedly in some of the world's largest, most complex organizations. I kept asking myself, "Why would the authors approach me to write this Foreword?".

To my delight, I found the book to be not just about Graphene as a material, but about how this amazing material could play a crucial role in the broader world of information technologies in the not-too-distant future.

The book was an eye-opener, an easy, breezy read. The way the authors structured the book is layered. They created the proverbial 'reader-hook' by framing a thriller-like multinational story of scientific discovery and development, spanning continents, leveraging exotic locales, peppered with intriguing characters, and set against a backdrop of high stakes and extreme urgency.

Set against this screen, graphene – as an amazing material, with pluripotent and patentable attributes, takes center stage.

Graphene is not a new material. Its base material, graphite, was first observed no doubt in ancient times as a substance that could mark surfaces. While that property may have been known, graphite was first discovered in its naturally occurring state, in bulk, near Borrowdale, England in 1564. Because graphite had a naturally occurring shiny sheen, it was first mistaken for lead. Borrowdale graphite was pure, solid, and easily cut. Literacy in the "modern world" was becoming democratized. Using this property of graphite as the basis, a new industry, of making "lead pencils" for writing was born.

Still erroneously called "lead," the humble central component of a "lead pencil" was taken for granted for the better part of the following four centuries. The term "graphite" was coined in the mid-1700s, by German chemist Abraham Gottlieb Werner, derived from the Greek work *graphein*, meaning "to write". It was only in the first half of the 20th century that interest in graphite was rekindled.

Graphene as a stand-alone material worth studying was first conceptualized in the 1940s. However, it was not until 2004 that it was successfully isolated and studied in a laboratory setting by Messrs. Andre Geim and Konstantin Novoselev at the University of Manchester in Britain. They isolated graphene in the most esoteric of ways; by peeling off layers of graphite with adhesive tape until they achieved a single layer of carbon atoms. Their work earned them the 2010 Nobel Prize in Physics. Subsequent studies have revealed the many different properties of this single-element material, made of just carbon atoms arranged in lattice structures.

Today, the widespread industrial use of graphene is limited by several challenges, despite its remarkable properties such as high strength, flexibility, conductivity, and thermal resistance. Structural integrity and purity are vital for graphene to extend all of its vaunted properties.

Production costs are still high, regardless of whether graphene is produced using chemical vapor deposition (CVD), exfoliation, or graphene oxide reduction. Producing high-quality and defect-free graphene in quantity is a significant barrier. Scaling to industrial quantities is proving to be a technical challenge. Graphene tends to 'isolate and stay aloof', making it difficult to integrate it with other materials and create adequate composites.

Market demand remains limited, as industries tend to be cautious about new materials until they are proven at scale and at a cost point that preserves or even enhances profit margins. Plus, there is concern about safety and environmental impacts.

Governments and companies are working together to overcome these barriers. Niche applications are emerging, and momentum is steadily building, potentially leading to breakthroughs in the near future. That could revolutionize numerous industries.

The Graphene Bridge center-stages the eponymously named material. It surrounds graphene with rich context, using Artificial Intelligence (AI), Quantum Computing, Atomic Batteries, and Cybersecurity. Using these four contextual components to frame an engaging story, the authors have stitched a tapestry of use cases.

The result is a fast-paced book, an industrial espionage thriller, set on an international stage. It provides an engaging look at the astounding qualities and use cases of graphene, the chief protagonist in this novella.

Give yourself the gift of a couple of hours. You will no doubt enjoy the book and come away with a much-improved understanding of ***Graphene*** as a material.

By Mr. Ashwin Rangan

Chief Executive Officer
DoubleCheck Solutions

Acknowledgments

We are indebted to Mr. Ashwin Rangan, Chief Executive Officer at DoubleCheck Solutions, for his insightful compelling foreword and executive review, which provides essential context while enhancing the book's overall narrative.

We are eternally thankful to Dr. Monica Cox, Distinguished Professor of Engineering at The Ohio State University and an eminent author, for her compelling afterword. Her depth of field is intriguing and bound to leave a lasting impression on our readers.

We are also grateful to the following individuals for their expert reviews, superb creative direction, mindshare, and timely advice, which significantly enriched this book:

Dr. Abhimanyu Gupta, Instructor, Department of Operations & IT Management at Richard A. Chaifetz School of Business, Saint Louis University.

Mrs. Kathyrn N. Guzner, Attorney and Founder of Guzner Law, P.C.

Mr. Kevin Kokinda, Resumedics Founder and Author of Finding Your Career Mojo: Muddling thru the Resume' BS.

Mr. Mark Settle, Former CIO and Author of Truth from the Valley, Truth from the Trenches.

Finally, we would like to express our heartfelt gratitude to our families and friends for their unwavering support and encouragement throughout this journey.

Prologue

"The best way to predict the future is to invent it." - Alan Kay

Picture this: You're playing chess against a computer that's not just better than you, it's better than every human who's ever lived, combined. What if you don't need to charge your electric car for hours, but seconds, and then drive for years without ever plugging in again? How about a material so thin you can't see it, yet it's stronger than steel and more conductive than copper?

As Arthur C. Clarke famously said, "Any sufficiently advanced technology is indistinguishable from magic." We stand on the brink of a revolution that will render today's technology primitive by comparison.

Welcome to the bleeding edge of innovation, where graphene is rewriting the rules of what's possible. This is not merely an incremental improvement or a nifty gadget; we're talking about a foundational shift that could reshape our entire civilization.

Graphene, a material thinner than paper, stronger than steel, and more conductive than copper, is poised to revolutionize numerous fields.

But here's the kicker: for all its mind-bending potential, graphene, like other cutting-edge technologies, presents both immense opportunities and significant challenges. They're like newborn gods, immensely powerful, but also unpredictable and in need of guidance.

Whether you're a tech enthusiast, a business leader, or simply someone trying to make sense of our rapidly changing world, understanding these frontiers is essential. The decisions we make now and the ways we choose to develop and deploy these technologies will shape the future for generations to come.

Each promising innovation, whether in graphene, artificial intelligence, quantum computing, or energy, is both an opportunity and a challenge. Like unpolished gems, these technologies require careful shaping to unlock their full potential. And that's where this book begins: a journey into the transformative power of graphene and its potential to shape our future. Welcome to the next chapter of innovation, where graphene is the bridge connecting what we know with what we can only imagine.

Introduction

A hypothetical story follows in *italics* ...

Berlin was alive with energy. Trams flickered through the neoclassical facades, the sound of their electric hum underscoring the pedestrians' conversation. But underneath all that glittering density, there was a calm before the storm.

Inside the sleek lab of the World Graphene Council (WGC), framed by glass panels, it yielded to tension that crackled with the promise of static electricity. Dr. Elena Petrova was standing in front of a vast spread of a holographic display, scanning torrents of streaming data as though she were some Prime Axiom Architect of the Graphene Bridge; an overwhelming pan-global cross quantum/ AI system would flood her graphene enhanced systems ... if it worked. And if it survived. "This is not some academic exercise," Elena's sharp voice cut through the room's stillness. "We're staking everything on this."

Anya Morgan, the lab's quantum computing specialist, stood leaning against a console, her sharp blue eyes flashing challenge and doubt. "And rewriting reality while we're at it," she clipped. Kai Mitchell, the atomic battery expert and eternal pragmatist, just unfolded his arms. "Reality's not the issue," he muttered. "Integration is. All the tech in the world; it's useless if it's not going to play together."

The lab door hissed open. Agent Reeves stepped inside, and at once, he seemed to change the room's whole energy. Never mind the chaos, he was in charge, every inch of him oozing from that tailored suit. They called him the prophet - having this amazing knack of foretelling catastrophe before it happened at all. "Progress attracts attention," Reeves began, his voice calm but commanding. "Novak's network is monitoring our advancements in the Graphene Bridge platform. We must assume they're planning to exploit vulnerabilities."

Elena's gaze sharpened. "What other kinds of threats?"

"Sabotage. Theft. Disinformation. You name it," Reeves said. "This is not just about innovation anymore. It's about survival."

Anya's lips twisted into a faint smirk. "No pressure," she quipped. "Just save the world and dodge rogue networks while we're at it."

Kai chuckled under his breath. "At least Novak's got good taste. He's going after the best tech out there."

"Focus," she snapped at them with a quick look. "If we're not precise, everything falls apart."

Her words' weight hung in the air between them, pressing into each other. Outside, the Spree River sparkled in the sun, dancing oblivious to the deep stakes being taken within this lab. Pedestrians rushed by, heads unbowed and unknowing that Armageddon was just a few blocks away. The team turned back to their stations. Anya futzed around with quantum encryption algorithms trying to perfect them to resist even a Novak-style blitzkrieg. Kai fine-tuned his atomic battery prototypes simultaneously pushing the envelope on the safety-technique threshold. And Elena race-modeled through graphene synthesis algorithms trying to close every orifice for an 'aha!' moment from Novak. They met again hours later. Holographic charts and models stood in the air, glowing symbols of months of inhuman exertion. Reeves stood at the edge of the room: a silent guardian. "Graphene is conductive," Anya said, beginning crisply. "That's the edge we've got. Novak's going to hit us."

"Atomic batteries-that's a game-changer," Kai said. "But they're junk if Novak hijacks our supply chain."

Elena nodded, her voice grim. "Every piece matters. If one goes falter, the whole Bridge goes down."

Reeves finally spoke, and it was in a low but forceful tone. "You have made something remarkable. And yet Novak is always counting on those 'cracks in the foundation.' I fear he may have already made his move."

Anya's smirk returned, slightly softer this time. "Good thing we're always one step ahead, huh?"

Nobody was laughing now. It was the huge size of their assignment that settled on them like a storm. Out in Berlin, life continued as it had for years, and not even the existence of even a subtle war was suspected around and within the very walls of the WGC. Inside, the team set their heads down and went about their work, resolved to maintain a line.

...

Most people grasp Artificial Intelligence (AI) to some degree, but quantum computing on the other hand may open some grey areas and even more when it comes to atomic batteries. For now, think of each of these three components in three separate buckets of paint: red for AI, blue for quantum computing, and green for atomic batteries. Of course, what other color would there be to represent batteries? The fun part will be at the end when we mix the colors to create a vibrant masterpiece.

AI functions as a machine brain. We associate it with red, though it's far from having a circulatory system[1]. You see, the most basic computer is simply a worker waiting to be told what to do. It needs orders and it's not going to think for itself. Whoever the chosen user is, must be knowledgeable about whatever the subject may be and know how to navigate a system, program, or even just a browser. It may surprise you how complex a web search can get when you're

[1] AI, short for Artificial Intelligence, mimics certain cognitive functions of human beings, such as learning and problem-solving.

looking for a very specific piece of information. If you think you've got that down, just read what Google dorking is[2]. However, AI is for all of you who don't want to keep giving instructions. AI is more like a concierge. You enter a marvelous, warmly lit lobby with magnificent artwork decorated along the walls and beneath one golden spotlight is a person who can figure out anything for you. They may not know immediately all the answers, but they're quite determined to figure it out. AI excels where a human falls short, particularly in quick information retrieval and vast memory storage. AI systems come in various forms, each with its unique way of learning, operating, and engaging. It's as if you have access to multiple concierges simultaneously, each with distinct characteristics and capabilities, rather than relying on a single staff member.

In the last two decades, three key areas have shaken up innovation in the digital world: Quantum Computing, Artificial Intelligence, and Atomic Batteries. Their convergence creates unprecedented opportunities, particularly when graphene bridges their respective limitations. These aren't theoretical concepts anymore, but rapidly evolving realities with the potential to reshape our world.

The science fiction of the past has metamorphosized into science fact.

---

[2] Advanced search techniques, such as 'Google Dorking', allow users to refine search queries with greater specificity.

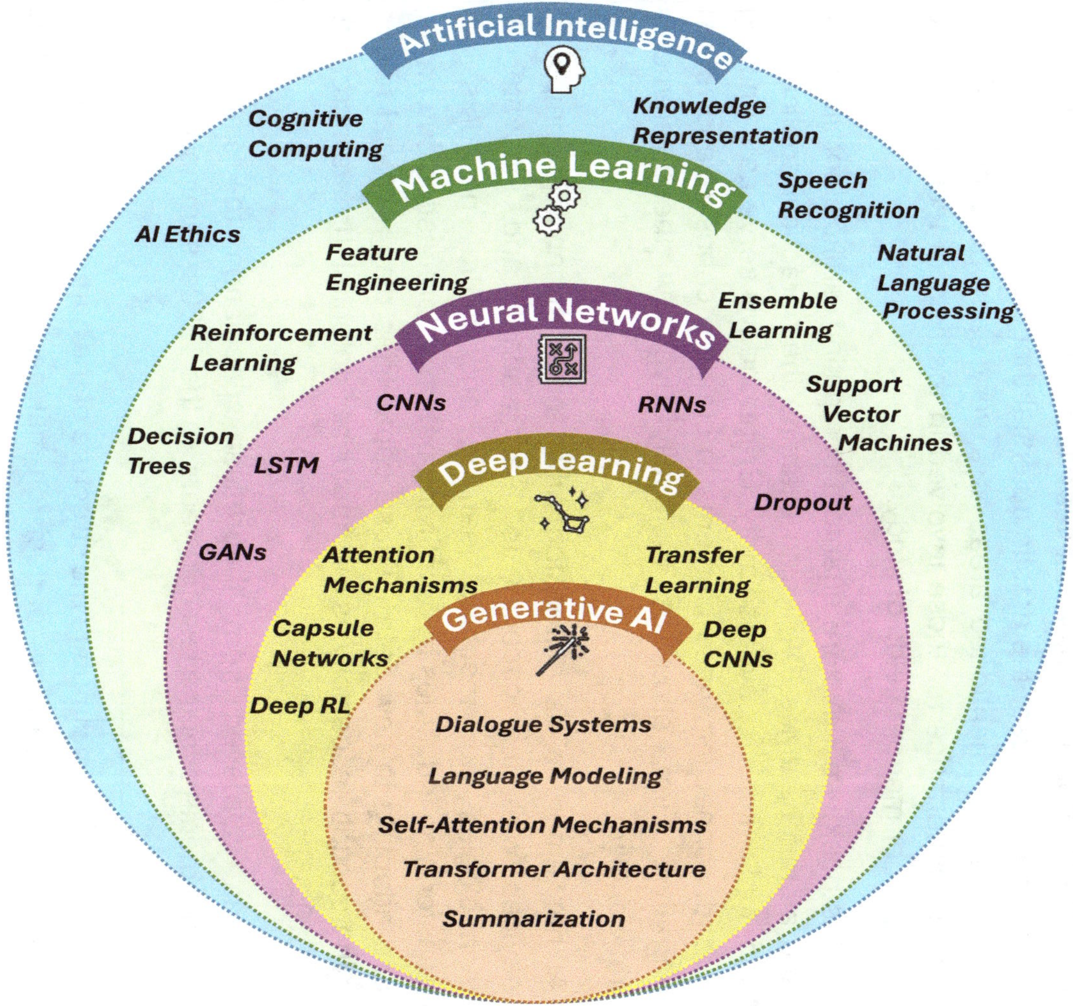

Figure 1: A layered taxonomy of Artificial Intelligence (AI), illustrating its components from Machine Learning and Neural Networks to Deep Learning and Generative AI. The visualization highlights AI's catalytic role in advanced decision-making that is crucial for tackling challenges like cybersecurity.

It is just the start of our journey into these game-changing technologies, which are all linked and set to transform the world [7]. Our objective is to open a space where major technologies such as these can be understood and bridged by a major element that can change the world in drastic scenarios: **graphene**. Regardless of any prior knowledge you may have heard of graphene or the more commonly known graphite that lies within the pencils that most of us do not regularly use, the drastic potential that graphene can have in the three aforementioned areas is a very early prospect with the utmost importance for those involved in the field of technology and many other modern fields of study.

In this book, we'll explore quantum computing, artificial intelligence, and atomic batteries. We'll also examine their practical applications and implications for cybersecurity. Graphene will serve as a unifying thread, connecting these elements to create a cohesive and robust narrative. First, allow us to briefly overview these elements individually.

AI has emerged as a transformative force, enhancing computing devices with the capacity to learn, adapt, and innovate. AI technologies like machine learning and neural networks are transforming industries by identifying patterns, forecasting outcomes, and adapting to new data[3]. These capabilities play a crucial role in cybersecurity, especially as the number of potential threats increases [16,19,35].

AI has journeyed from theoretical speculation to remarkable achievements over the last 100 years. In the 1950s, seminal concepts

---

[3] Neural networks simulate the structure of the human brain to process data and create patterns for decision-making. They're used to analyze large datasets, make predictions, and adapt to evolving information.

like the Turing Test and early AI programs laid the groundwork for future exploration [16].

The Dartmouth Conference in 1956 marked the official commencement of AI research, leading to breakthroughs such as the GPS algorithm and the Lisp programming language in the following decades[4]. Despite funding setbacks in the 1970s and 1980s, interest in AI persisted, culminating in the 1990s with the resurgence sparked by deep learning techniques [16].

Fast-forward to today, AI stands as a testament to human ingenuity and technological prowess. Since the 1990s, monumental victories such as IBM's Deep Blue defeating chess champion Garry Kasparov in 1997 and the triumph of IBM's Watson over Jeopardy champions in 2011, have underscored AI's capabilities [19]. This journey reflects not only the evolution of technology but also the relentless pursuit of understanding human cognition and AI via complex computing algorithms and systems.

AI algorithms are trained on massive datasets, allowing them to identify patterns and make predictions with superhuman accuracy. These datasets can include anything from medical scans to financial records to social media posts [19]. AI digs into huge amounts of data to find hidden gems of information and make predictions that humans couldn't manage on their own. In medicine, for instance, AI can study medical scans with pinpoint accuracy, helping spot diseases early and tailor treatment plans to individuals. For example, let's consider an increasingly common scenario. You're planning a birthday party with the assistance of a voice-assisted planner on your smartphone.

You begin with the prompt “Siri” or “Hey Google”, which jumpstarts the AI-driven event planner. This digital ally analyzes your past celebrations, breaks down preferences, and offers personalized suggestions for the perfect venue, decorations, and even

[4] The Dartmouth Conference is widely regarded as the birth of AI as a field of academic study.

entertainment. Much like having a savvy friend who anticipates your needs, it can book venues on your behalf and look at availability and other helpful criteria such as cost while enhancing the planning process.

AI offers a future where machines blend effortlessly into our daily routines, understanding our preferences and serving as smart helpers[5]. Picture a home that not only adjusts temperature, lighting, and entertainment to your preferences, but it also predicts your needs based on your daily routines, health data, and emotional state. It could prepare a personalized meal plan, order groceries, and even synthesize custom supplements tailored to your unique biometric profile. These are just a taste of how AI could revolutionize our lives and make our lives easier.

Beyond these examples, AI has subfields and a plethora of applications, holding the potential to revolutionize personal interactions with technology. The shakeup in AI-driven technologies has been happening steadily over the past decade and the urgency for AI to solve more pressing issues has increased tenfold. The real potential goes beyond automation; it's about tailoring experiences to everyone, making technology a seamless part of our everyday routines. The growing prevalence of AI brings imperative ethical questions to the forefront. Key concerns include algorithmic bias, job displacement due to automation, and the potential misuse of highly capable AI systems.

As we move forward, it's crucial to develop and use AI responsibly. Our goal should be to harness AI's potential to benefit humanity and create a future full of opportunities. This optimistic view aligns with the vision presented in Diamandis and Kotler's book "Abundance: The Future Is Better Than You Think."

Quantum Computing

---

[5] AI is increasingly being integrated into daily life through smart devices, from virtual assistants to home automation systems.

Built upon the complex foundation of quantum mechanics, Quantum Computing is the avant-garde of cutting-edge computation, completely transforming our approach to solving problems [8]. Its ability to tackle complex issues much faster than traditional computers is truly game-changing. Imagine a world where complex calculations that would take classical computers centuries would take quantum computing mere seconds [8]. In recent years, like many pioneers in the field, Google has advanced its Quantum Computing efforts that have culminated in the Sycamore Quantum computer [21]. In 2023, Google compared Sycamore (with 72 qubits at the time) to the then-most powerful supercomputer, Frontier. Sycamore completed a specific calculation in seconds, whereas researchers estimated Frontier would take approximately 47 years [31]. It's important to note that the specific calculation Sycamore tackled was designed to showcase its quantum advantage.

Hence, real-world applications may require different algorithms and thus may not see such a dramatic speedup. Yet, this is what Quantum Computing offers: a world of possibilities that reach way beyond mere scientific interest. For instance, discovering new drugs by swiftly analyzing countless molecular structures, is like finding a needle in a haystack, but this needle could be the cure for many deadly diseases. Previously production-heavy tasks such as new material creation with properties never even dreamed of, like a super-lightweight structure with the same strength as diamonds are also a growing reality. Quantum Computing's impact extends beyond science as it could truly shake up industries such as medicine, logistics, and finance, to name a few.

Fast forward to 2025, Microsoft unveiled Majorana 1, a quantum chip leveraging its new Topological Core architecture. This breakthrough has been built on topoconductor materials (viz. special materials that are not a solid, liquid, or gas) and enables more stable and scalable qubits, potentially allowing for one million qubits on a single chip. By demonstrating the creation and measurement of Majorana particles[6], Microsoft's approach could accelerate the commercialization of quantum technology, an improvement in fault-tolerant quantum computing with wide benefits for multiple fields.

To unravel the profound implications of these concepts and lay the groundwork for understanding the interconnected world of Quantum Computing and its symbiotic relationship with other technological domains, let's put quantum computing in layman's terms. Take an everyday task like grocery shopping. A classical computer thinks in bits (as in flipping a coin for heads or tails) to figure out which aisles to shop in. A quantum computer uses its qubits to *simultaneously* explore *all* the aisles, thereby making problem-solving quicker, more efficient, and desirable in more walks of life. Imagine the phenomenal real-world implications of scaled multitasking.

Atomic Batteries

Stars illuminate the night sky with enduring energy and may hold untapped reserves for our future [5]. Similarly, Atomic Batteries have emerged as silent contributors that could revolutionize sustainable energy consumption.

Modeled on nuclear batteries but with synthetic materials, these innovative power sources promise unparalleled efficiency and sustainability. They fundamentally promise significantly longer lifespans and solid efficiency. Consider the critical roles pacemakers play in regulating heartbeat and how accurately they must function. Such devices powered by an Atomic Battery that lasts a lifetime, eliminate the need for risky and intrusive surgeries.

Atomic batteries introduce a new era where the smallest particles command transformative power, challenging our understanding of energy dynamics [5,20]. This technology could lead to a substantial

[6] Majorana particles are exotic quantum particles that, when hosted in special materials called topoconductors, can be used to create more stable and reliable qubits.

increase in the number of IoT devices, which, in turn, would enhance security systems across various domains. The power source in Atomic Batteries comes from radioactive decay that provides a consistent source of energy. Certain isotopes, unstable forms of elements, naturally emit radiation as they decay into more stable forms[7]. In an Atomic Battery, this radioactive decay is harnessed to generate a steady stream of electricity. The specific materials and isotopes used determine the power output and lifespan of the battery.

Coming to practical applications, envision a future where electric vehicles are powered not solely by conventional Lithium-ion batteries but by Atomic Batteries. In these trailblazing scenarios, atomic processes operate within the battery, utilizing synthetic materials to generate a sustainable and efficient energy source [20]. Imagine an electric car powered by atomic batteries. Unlike current batteries that need frequent charging, these could provide a steady, long-lasting energy supply. Picture this: instead of spending 3 hours to fully charge your car, you could recharge in just 5 minutes.

Not only that, but your car could then travel up to 2,050 miles on a single charge - that's 5 times the typical range of today's electric vehicles. And this impressive performance wouldn't be affected by external factors like cold weather[8]. This not only reduces the need for frequent recharging but also contributes significantly to a greener and more reliable transportation landscape. While the concept of Atomic Batteries is captivating, significant challenges remain. Safety is paramount, as these batteries contain radioactive materials. Additionally, managing the heat generated by the decay process is crucial.

Even with these challenges, progress in fields such as choosing the right radioisotopes, developing containment materials, and managing

---

[7] Common isotopes used in atomic batteries include Strontium-90 and Polonium-210.

[8] Atomic batteries have the advantage of providing consistent performance across a wide range of temperatures.

heat is paying off. Researchers are also working hard on making Atomic Batteries smaller, aiming to power things such as pacemakers and other implantable medical devices. Once just a far-off idea, Atomic Batteries are now becoming something we can actually hold in our hands. Their potential applications extend far beyond pacemakers, with possibilities for powering remote sensors, satellites, and even spacecraft on extended missions. Think about this: on a faraway planet, there are sensors powered by Atomic Batteries that can last for decades, beaming back important information to Earth. And, as we keep working on them, Atomic Batteries could lead us to a cleaner, more sustainable energy future, breaking free from the constraints of regular batteries.

Unlocking The High Potential of Graphene

Let's further extend the premise of primary graphene uses. As this thin material can be used in construction and surgery, it may also be used in CPUs and perhaps even replace cellphone antennas.

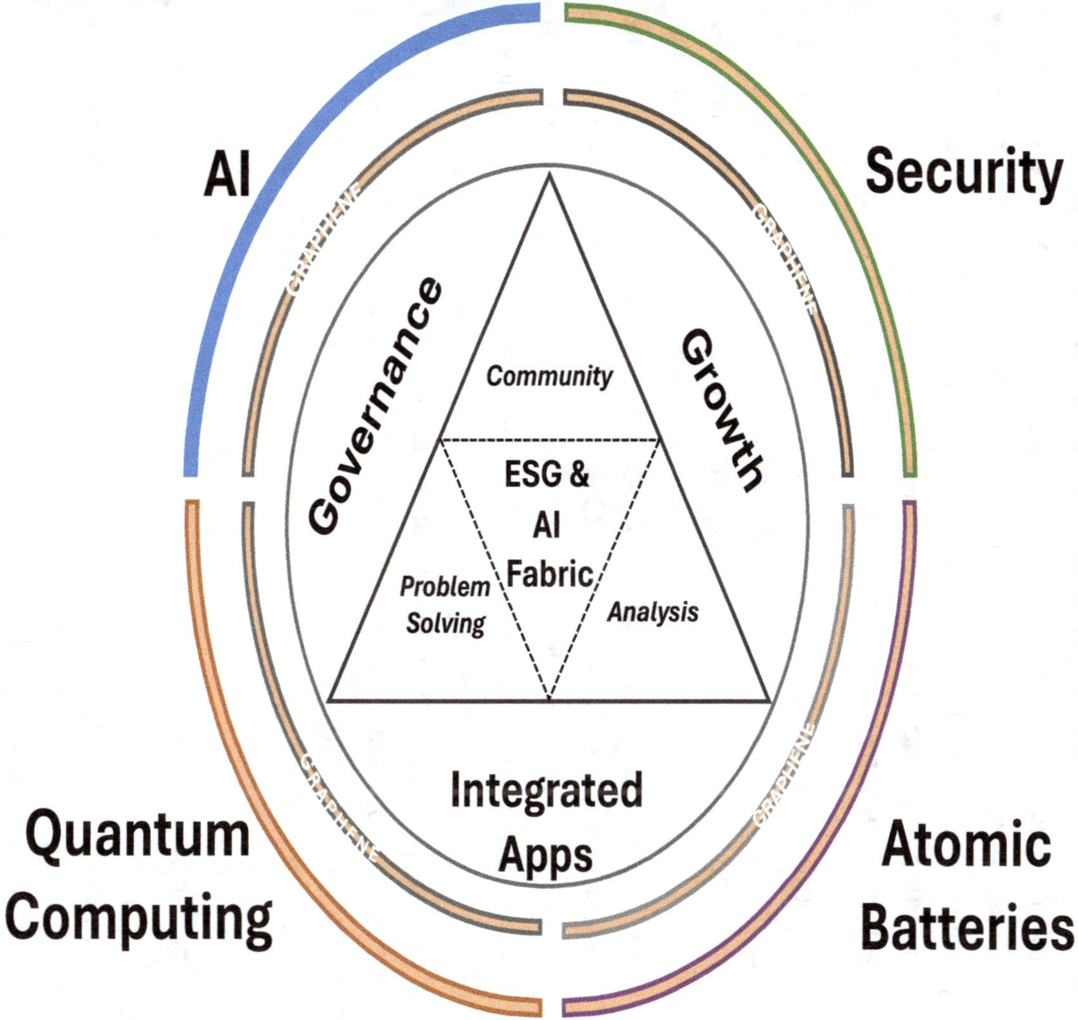

Figure 2: An integrated framework highlighting the synergy between breakthrough technologies: graphene, atomic batteries, and quantum computing systems.

Since it is stronger than steel and capable of conducting electricity faster than anything known to date, we'll examine graphene's history and structure in more detail. Born from the humble graphite in pencil lead, graphene has the potential to reshape fields like security, computing, energy storage, and more.

Graphene is a single layer of carbon atoms arranged in a hexagonal lattice, a structure that gives it unprecedented potential and extraordinary properties. It is 100 times stronger than steel, incredibly flexible, and an exceptional conductor of electricity. Unlike raw graphite, refined graphene exhibits unique electrical and thermal properties, making it a game-changing material for applications in electronics, energy use, and advanced materials science.

The journey to unlock graphene's potential is as compelling as the material itself. In the 1940s, scientists theorized the existence of a "zero-gap semiconductor" with unparalleled properties. Yet, for decades, graphene remained an elusive dream, its potential untapped due to the challenges of isolating it. The breakthrough came in 2004 when two researchers, Geim and Novoselov, using simple scotch tape, extracted single atomic layers of graphene from graphite. This achievement later awarded them the Nobel Prize in Physics in 2010, sparked a surge of global interest in the material, and marked a pivotal moment in its history. Within 8 years, more than 5 thousand academic papers detailed applications ranging from flexible electronics to energy storage. Graphene's strength, 100 to 150 times greater than steel by weight, makes it ideal for aircraft components, as shown in Table 1.

Historical Evolution

Graphene's journey is also a story of overcoming limitations. Its exceptional strength-to-weight ratio, 150 times stronger than steel yet remarkably lightweight. This has made it a top candidate for everything from Kevlar vests to "graphene concrete," or concrene, a composite that enhances the strength, stiffness, and water resistance of traditional concrete. Its unmatched electrical conductivity surpasses copper, making it a potential cornerstone of future battery technologies with lightning-fast charging times. Meanwhile, researchers are exploring its ability to regulate thermal energy, potentially enabling breakthroughs in solar heating, cooling, and alternative energy systems. Nevertheless, widespread adoption depends on addressing challenges such as material consistency and integration with legacy systems.

Graphene's most exciting potential lies in enabling room-temperature superconductivity, removing costly cooling needs, and transforming energy storage and transmission [21,25]. This would be a paradigm shift, opening doors to advancements in AI, quantum computing, and atomic batteries, technologies that demand materials capable of bridging performance gaps at unprecedented levels.

Despite its promise, graphene faces significant obstacles. Its lack of an intrinsic off-switch in electronic circuits complicates its use in transistors. Scaling up high-quality graphene production is both costly and technically challenging. Current methods, such as mechanical exfoliation and chemical vapor deposition, remain expensive and difficult to implement at scale. More sustainable approaches, such as using carbon dioxide as a carbon source, offer hope for cost-effective and scalable production, but these techniques are still under development.

The cost of high-quality graphene, currently up to $200 per gram underscores the need for further innovation in production methods. However, researchers remain optimistic. The extraordinary combination of properties graphene offers makes it a crucial material for the future, and ongoing advancements are expected to bring down costs and overcome scalability challenges.

Table 1: A summary of graphene's key properties and their implications for various applications in multiple domains.

Property	Description	Implications for Applications
Strength	100-150 times stronger than steel by weight	Lightweight, durable materials (ex: aircraft components, Kevlar alternatives, and structural reinforcements)
Flexibility	Stretchable up to 150% of its original length	Flexible electronics, medical implants, and wearable devices
Electrical Conductivity	Conducts electricity faster than copper and silicon	High-speed transistors, advanced batteries, and highly sensitive sensors
Thermal Conductivity	Exceptional thermal regulation properties	Cooling systems for high-performance electronics and thermal energy applications
Transparency	Nearly transparent, with a single-atom thickness	Transparent touchscreens, lightweight solar panels, and augmented reality lenses
Elasticity	Maintains structure and elasticity even when stretched	Elastic composites, deformable medical implants, and stretchable circuits
Chemical Stability	Resistant to chemical degradation	Longevity in harsh environments for membranes, industrial coatings, and reactive sensors
Light Weight	High strength-to-weight ratio	Aerospace and automotive components that demand reduced weight and high durability
Impermeability	Blocks all gases, including helium	Gas filtration systems, leak-proof storage solutions, and protective layers for sensitive electronics

The journey of graphene, from its humble origins in graphite to its position at the forefront of technological innovation, is a testament to human ingenuity. As we continue to unlock its potential, graphene holds the promise of bridging today's challenges and tomorrow's possibilities, a material poised to shape the future of our world.

PART 1

Artificial Intelligence & Quantum Computing

Chapter 1

Graphene-Based Sensors, ML, Neural Networks

London wasn't Berlin. While Berlin had been the cradle of their theoretical advancements, London's cutting-edge labs offered the infrastructure to test the Graphene Bridge in action. The relocation had promised improved labs, cutting-edge tech, and finally, the ability to bring the Graphene Bridge to life. Yet, all it delivered was a further knockback in trying to roll out graphene-enhanced sensors intertwined with machine learning and neural networks.

"Damn," Anya Morgan muttered through clenched teeth, her fingers a blur across her console. The neural networks were learning way faster than anybody had projected; too fast. The graphene sensors were driving their algorithms toward the breaking point, and everyone knew it.

"Battery status?" still without looking at the readout, Kai Mitchell shrugged over one of the test cells, his usual smile absent. "They're dying in the humidity. One rainstorm and we're well and truly dead in the water."

A shrill wail of alarm cut through the tension. Anya's screen flashed. "Sensor overload!" Desperately, she began to hit the keyboard, but it was already too late. A skein of smoke wrote lazily up from the prototype, carrying the acrid tang of burned possibilities and wasted time. "Brilliant," Kai grumbled, "You've outdone yourself, again."

The lab doors hissed open. Major General Sir David Blackwood was filling the door like a thundercloud, his face even darker than the smoke, wisping still from their failed experiment. "Tell me," he said, each word sharp as a piece of broken glass, "that this is not what I am going to be seeing at next week's defense demonstration." Elena held his gaze.

"Innovation isn't a straight line, Sir." "Neither is unemployment, Dr. Petrova." Reeves appeared as if from nowhere, his presence adding another layer of tension to the charged atmosphere. 'Novak's network is watching closely. There was a data breach attempt last night targeting your supply chain reports,' Reeves warned. "Are they in?" "No." Reeves' face lines.

"But I guess they know where to look exactly." The implications hit Elena like a physical blow. "They knew about the sensors." She leaned close so that her lips almost touched his ear. "Novak's got people watching us." 'Did they get through?' 'Not this time,' Reeves replied. 'But whoever's behind it knows exactly where to hit.'

The crew broke to their positions after the tactical huddle, yet impotence haunted them all. Every keystroke, tweak, and test subjected them to the agonizing awareness of eyes they could not see. By sundown, they had something that was a working prototype-just. It hummed weakly on the bench, its sensors flickering blue in what felt like a gloomy premonition. "It's alive," Kai smiled wanly at his joke. Anya hunched over her station, dark smudges under her eyes as she fine-tuned the error protocols.

"These things need to think for themselves," she grumbled.
"But not too much." Elena studied the holographic Bridge model rotating gently above the bench site; every part of it was shown in blue ghostly form. "One weak link, and the whole thing fails," she said. "And Novak gets exactly what he wants," Reeves added from his corner. They worked in silence, each lost in their own thoughts.

The Graphene Bridge was no longer a project. It was a race. A weapon. A target.
And they were running out of time.

...

Graphene's exceptional electrical properties, high surface area, and mechanical flexibility make it an ideal material for developing advanced electrical sensors [10]. When integrated with Machine Learning (ML) and Neural Networks (NN), these next-generation sensors also have the potential to transform medical applications. From material and technological evolution to the graphene revolution, this wonder material is where innovation is meeting practicality. Let's examine how graphene is intertwined with electrical sensors, ML, and NN to develop groundbreaking applications for mankind.

Graphene-based Sensors

Sensors are integral components of machines and devices that measure readings and play an important part in our connected world. However, certain limitations in sensitivity, speed, and operational requirements often hinder the full potential of traditional sensors. Graphene showcases great electrical conductivity, high surface area, and amazing mechanical flexibility, and continues to revolutionize the sensor technology field. Researchers at Columbia University have engineered ultra-sensitive air quality sensors utilizing graphene that can detect minute traces of pollutants like nitrogen dioxide [5,20][9].

These integrated graphene-based sensor technologies have the added benefit of being deployable in dense urban environments to provide real-time air quality data, which is crucial for public health initiatives and environmental regulations. Graphene-based sensors are here to sustain our most advanced needs and future-proof this sub-market for years to come.

[9] Graphene's atomic thickness and conductivity make it ideal for detecting pollutants in real time.

Graphene's ability to interact with microscopic biomolecules makes it ideal for next-generation (NextGen) sensors [21]. These combinations of materials and technologies surpass the capabilities of traditional sensors in terms of sensitivity, speed, and operational efficiency. For instance, researchers at the University of California have successfully created a graphene-based NextGen sensor for detecting dopamine, a neurotransmitter with significant brain functions [18]. Naturally, this biosensor can play an integral role in neurological research and the development of viable therapies for afflictions such as Parkinson's disease. As seen in Table 1, graphene's flexibility, stretchable up to 150% of its original length, enables its use in wearable devices and biosensors. In light of these capabilities, leading electronic manufacturers such as Samsung have invested millions of dollars into R&D for wearable technologies. They're actively exploring graphene-based sensors for smartwatches to monitor core biometrics like heart rate, blood pressure, and blood glucose levels with greater accuracy and comfort. This new wave of capabilities and scientific possibilities can see hospitals, clinics, physicians, and everyday individuals taking a more proactive approach to health management and wellness [7]. This small set of combined technologies has the potential to quickly transform our interactions with the world around us.

Considering the significant improvements that graphene may have on sensors, especially regarding the collection, transmission and unification of sensitive data, it is paramount that the security of that data remains addressed. Since the world of components is on a steep trend towards convenience and mobility, many of these sensors include internet capabilities which groups them into the vast world of the Internet of Things (IoT). We hope that you are well versed of the dangers such as insecure IoT cameras being exploited by countless websites and broadcast to the general public. If you have a camera that connects to the internet, you may be a victim as well. Security precautions should be at the forefront of our minds right alongside the innovations themselves.

Data encryption, multilayer security, and firmware/software best practices leads to making the sensors and IoT devices more secure. However, these advancements must address security challenges, including device spoofing and unauthorized access. Data encryption can be drastically improved by implementing a virtual private network (VPN). Creating a tunnel between the device (ex: a dopamine reader) and its endpoint (sensor or IoT device) enhances data security by making it significantly harder to intercept and decrypt. Different algorithms for data encryption within these VPNs, like all other technologies, are constantly changing and improving.

However, firmware and software for these sensors within the domain of IoT is a bit more difficult to manipulate considering there are less variables to take control of. Oftentimes, these devices maintain a firmware that becomes outdated and accrues many methods of exploitation. The same case is for their larger counterparts, software, which interacts with a middleman that speaks to the hardware as opposed to directly speaking to the hardware as firmware does. Since software has a bit more opportunity for user input, it is important to ensure that privileges are granted on an as needed basis. Oftentimes, IoT grants full permission as a default, offering more information that a potential threat actor could get their hands on. So, turn off those readings within things like graphene-based smartwatches when you are not interested in having that information at the moment.

Machine Learning (ML)

What is Machine Learning or simply put, ML? It is a subdomain of Artificial Intelligence [16]. Just as the human brain learns new concepts and adapts over time, ML allows computers to achieve similar results. By ingesting voluminous data, ML algorithms are often part of modular ML models that can identify patterns, make predictions, and improve their performance without specific programming needs on the exact outcomes [19]. However, the quality of this data plays a crucial role in this process. This is where graphene-based sensors come into play, poised to revolutionize the field of ML.

As we've previously discussed, graphene sensors surpass traditional ones in data acquisition speed and sensitivity, capturing intricate details at remarkable rates [10]. This specific quality primarily translates to richer datasets for training ML models. High-quality data is the backbone of effective machine learning; the richer and more detailed the data, the more accurately the model can learn and make predictions [19]. As seen in Table 2, supervised learning algorithms, essential for predictive models, rely heavily on high-quality datasets from graphene sensors. For instance, researchers at MIT are developing graphene-based pressure sensors capable of detecting incredibly delicate changes in pressure. Imagine a wide network of these sensors embedded in roadways. They could capture detailed, real-time data on traffic flow, road surface conditions, and even minor changes in pressure that signal the formation of potholes or cracks. This high-fidelity data can be fed into ML models to predict traffic patterns, optimize traffic flow in real-time, and enhance road safety by identifying potential hazards before they become critical [19,20].

By providing ML models with richer, more accurate datasets, graphene-based sensors can dramatically improve the accuracy and reliability of predictions. Furthermore, these advancements support the development of self-driving vehicles, allowing for safer integration of autonomous and non-autonomous vehicles on shared roadways. These not only improve safety measures but also reduce congestion on the roads. As we have been observing, graphene-based sensors enhance the quality and granularity of training data available for ML models. This improvement in data quality directly leads to better-performing, more reliable ML applications, revolutionizing how we approach problems like traffic management and road safety.

There are important subfields of ML in the context of graphene integration that we'll soon explore in the neural networks section. Meanwhile, just as ML is a subdomain of AI, Deep Learning (DL) is a powerful subset of ML that utilizes artificial neural networks with multiple layers to process complex data. However, unlike traditional ML algorithms that can function with smaller datasets, DL models thrive on vast amounts of data (refer to Table 2). This is where graphene-based sensors come into play. Where ML usually functions with a lot of supervised learning models, DL thrives in unsupervised learning use cases where outliers and new data may present opportunities for advanced decisioning. For instance, researchers at MIT are developing graphene-based pressure sensors capable of detecting incredibly delicate changes in pressure. A wide network of these sensors embedded in industrial machinery could revolutionize maintenance and safety protocols. The raw data from sensors can be fed into AI models to predict equipment wear and tear, monitor operational efficiency, and even pair it with environmental data to identify potential issues before they become critical, ensuring optimal performance and reducing downtime.

In the medical field, graphene sensors are already being explored for continuous blood glucose monitoring[10]. These sensors can generate a holistic, comprehensive data stream that ML algorithms can also analyze in real time. For starters, it is revolutionary for the treatment and management of diabetes, allowing for better monitoring and personalized insulin delivery, ultimately improving the quality of healthcare and life for diabetic patients. The affinity between graphene sensors and ML also has the potential to galvanize many industries, from environmental monitoring, defense, and automotive, to industrial automation, paving the way for a future driven by intelligent data analysis.

Neural Networks (NN)

Neural Networks, or NNs, are inspired by the intricate structure and complex functions handled by the human brain. They are a powerful tool in the AI arsenal [16]. Much like Machine Learning models, NNs are designed to learn and adapt over time, mimicking how our brains process information. However, as with any powerful technology, there are important security considerations to keep in mind.

Malicious actors could potentially misuse NNs for harmful purposes. For example, there have been instances where AI systems were manipulated to spread misinformation or generate deepfake videos, leading to significant societal harm. To prevent this, robust security measures need to be implemented. Additionally, NNs trained on incorrect or biased data could generate misleading information. A notable example is when an AI-based recruitment tool developed by a major tech company was found to be biased against female candidates, as it had been trained on resumes predominantly from men. This underscores the importance of vigilance.

---

[10] Graphene sensors are ideal for continuous blood glucose monitoring due to their sensitivity and high signal quality.

Sensitive data should also be carefully managed to avoid exposing critical information. For instance, a healthcare AI system that inadvertently exposes patient records due to poor data handling practices could put both companies and their customers at significant risk.

Despite these considerations, NNs hold immense potential, particularly when supported by high-quality data. Much like the human brain, NNs rely heavily on rich, accurate data to function effectively [16]. This is where the capabilities of graphene-based sensors come into play. As previously discussed, these sensors provide exceptionally detailed and high-fidelity data, which is crucial for training foundational models. For instance, graphene-based sensors can capture intricate environmental details at extraordinary rates [10]. Imagine a network of these sensors deployed in a smart city environment, monitoring air quality, noise levels, and other critical parameters in real-time. The rich datasets provided by these sensors are invaluable for building robust models. For instance, the datasets can be analyzed by NNs to predict pollution trends, optimize traffic management, and even improve public safety by identifying areas with high noise pollution or poor air quality [19].

To be clear, while there are critical security considerations to address when utilizing NNs, the integration of high-quality data from graphene-based sensors can significantly enhance their capabilities. This combination of advanced sensing technology and neural network modeling has the potential to revolutionize various fields, driving innovation and improving outcomes in ways previously thought impossible.

NNs make up the core of AI, and they rely on a plethora of learning models. As you can imagine, deep learning models require immense computational power. Graphics Processing Units or GPUs are computer components that evolved from gaming circuitry in game consoles such as the Sega Genesis and PCs to handle these massive workloads efficiently. To assist, AI frameworks like TensorFlow and PyTorch leverage GPUs to accelerate the training process for NNs.

While we won't dive into the complexity of these frameworks, it's worth noting that traditional ML algorithms require vast amounts of labeled data for training. The good news however is that new learning methods have been designed to help AI systems learn more efficiently, using much less data than traditional approaches. Here's what each of the new learning approaches means:

- **Zero-Shot Learning:** The model learns new concepts without any labeled data for the new class. Imagine teaching a computer to recognize animals it has never seen before. With zero-shot learning, the AI can identify a zebra without ever having seen a picture of one. It does this by using information it already knows about other animals and applying that knowledge to figure out what a zebra might look like.

- **One-Shot Learning:** The model learns a new concept from a single labeled example. This is like showing the AI a single picture of a zebra and saying, "This is a zebra." From just that one example, the AI can then recognize other zebras in different pictures. It's similar to how a child might learn to identify an animal after seeing it just once in a book.

- **Few-Shot Learning:** The model learns a new concept from a small set of labeled examples. This method involves showing the AI a small handful of examples - maybe 3 to 5 pictures of zebras. From these few examples, the AI learns to recognize zebras in general. It's like giving a quick introduction to a new topic before expecting someone to understand and apply that knowledge.

These methods allow AI to learn more like humans do - quickly and from limited information - rather than needing thousands of examples to understand a new concept.

LangChain and Chain-of-Thought (CoT) Programming

LangChain is a fairly new, emerging paradigm that focuses on programming languages that enable AI systems to reason and chain their thoughts, closely mimicking human-like explanations for their actions.

While still under development, Chain of Thought (CoT) programming holds high promise for more interpretable and trustworthy AI. CoT is an approach to help AI models think more like humans when solving problems. Instead of just giving a final answer, the AI is encouraged to show its work, explaining each step of its reasoning process. Imagine you're solving a math problem. You don't just write down the final answer; you show your work, step-by-step. CoT programming teaches AI to do the same thing. This method often leads to more accurate results, especially for complex tasks that require multiple steps of reasoning.

By combining these approaches, developers can create more powerful and transparent AI applications that can tackle complex tasks while providing clear explanations of their thought processes.

Graphene sensors, with their unmatched high sensitivity and ability to capture minute details, can provide the rich data that NNs need for computation. An apt use case is when paired with blockchain for decentralized authentication, graphene-based IoT systems can mitigate risks effectively. This also opens doors for groundbreaking biomedical applications, like brain-computer interfaces (BCIs) utilizing graphene electrodes as demonstrated by researchers in the University of California. Graphene-based electrodes can capture brain activity with exceptional detail, allowing NNs to decode these complex neural signals. The potential for improving quality of life and human-machine interaction through BCIs shows high promise. A future where individuals with paralysis can control prosthetic limbs or communication devices directly through their thoughts is tenable and well within reach.

The increasing synergy between graphene, sensors, electrodes, and NNs extends beyond prosthetics, holding vast potential for the robotics field. Robots equipped with graphene sensors can gather real-time data about their environment. This data could be augmented based on the use case and can be live-streamed or fed into NNs, allowing them to make split-second decisions and navigate complex situations with human-like dexterity. This paves the way for robots to play a more significant role in various tasks, from search and rescue operations, and hazmat auto exploration, to intricate surgeries. Sci-fi situations like robots collaborating with surgeons in extremely delicate procedures or assisting with search and rescue missions in disaster zones are closer to becoming reality. Thanks to Graphene technology integration, the possibilities for pushing the boundaries of automation covering machine-to-machine through business-to-business and human-to-machine collaboration are truly exciting.

Highly intelligent machines can now be shaped using a bedrock of graphene-based sensors and NNs integration. These machines will not only be able to quickly learn and adapt but also actively engage with the world in increasingly sophisticated ways as we step into a connected future.

Considering potential impacts that neural networks may have, risk mitigation and secure procedures have become a global concern amongst nations as a collective. Sam Altman, the CEO of OpenAI and developer of the well-renowned ChatGPT called for a global meeting in 2023 to discuss the security concerns associated with non-governed AI tools. The result of this meeting was an acknowledgment of just how dire and urgent these concerns are. Altman also slowed the consequent forecasted release version of ChatGPT (4) to better address these concerns in spite of the rapid growth and estimated revenue of over $1.6 billion annually that it was generating [34]. With established (additional) players like Google's Gemini and emerging contenders like DeepSeek AI advancing the landscape, the need for responsible AI governance and energy-efficient deployments is more urgent than ever.

From the perspective of users, there are a few concerns to be aware of including malignant usage, misleading information, fabricated information, inadvertent disclosure of sensitive information, and information security. Considering these models are constantly learning, it is very possible to feed them incorrect and possibly malignant data to get these to generate incorrect results, outputs or suggestions, this phenomenon is called data poisoning.

Table 2: A heatmap categorizing AI, ML, and NN, their integration with graphene sensor technology, and their alignment with Chain-of-Thought hierarchy components.

Category	Description	AI Focus	Graphene Sensor Connection	CoT Hierarchy Component
Artificial Intelligence	Provides context and goals for ML algorithms, defines overall system goals for sensor data analysis.	High	Medium	Prompting (Guiding ML models with structured input, enabling them to generate context-aware responses)
Machine Learning (including deep learning)	Utilizes data from graphene-based sensors to improve performance continuously.	Medium	High	Self-Consistency (Ensures that ML outputs remain uniform and logical across scenarios, enhancing reliability.)
Neural Network (NN)	May analyze complex patterns in Graphene sensor data to support system goals defined by AI.	Low	Medium	ReAct (Reasoning & Acting)

As a result, a user may request various metrics such as dates of a historical event and the model may refer to an incorrect date. Misleading information can undermine trust in the system and lead to poor decision-making if not properly managed and verified. Hence, it is crucial to implement robust but secure mechanisms to ensure the accuracy and integrity of the underlying data used by these models [35].

ML models, including large language models, present significant challenges when relied upon as sole sources of information, particularly for topics unfamiliar to users. These models are fundamentally constrained by their training data, which has inherent limitations in scope and currency. They lack genuine comprehension and cannot produce original research or generate new knowledge, often struggling with complex reasoning or nuanced subjects beyond their training parameters [19].

The issue of bias is a critical concern in these systems. The training data can mirror societal prejudices in human-generated content, and underrepresenting certain groups or perspectives can lead to skewed outputs. Moreover, the framing of queries can influence model responses due to learned associations, potentially reinforcing existing biases [4].

The heavy dependence on human input in AI training is another key factor[11]. The quality and accuracy of model outputs are directly tied to the caliber of human-curated training data.

Human biases and errors in data collection, cleaning, and labeling can propagate through the model, potentially perpetuating existing misconceptions or knowledge gaps present in the training data.

Given these considerations, users must approach AI-generated information with critical thinking. Cross-referencing information from multiple reliable sources, recognizing that models reflect human knowledge and biases rather than objective truth, and using AI as a

[11] AI systems are influenced by human-curated training data, which introduces biases.

complementary tool to human expertise rather than a replacement are crucial practices.

While language model creators have implemented some controls into their models to prevent users from getting assistance in engaging in suspicious and illegal acts, there also lie vulnerabilities to maneuver around these restrictions. There also lies an information security best practice to not expose sensitive data to the model.

Graphene-based sensors also play a vital, dual role in AI model risks. Their high-fidelity data can inadvertently amplify risks like data poisoning if malicious actors introduce corrupted inputs. However, these sensors can also mitigate risks by ensuring consistent and reliable data quality, enabling dynamic validation to detect anomalies, and enhancing data integrity protocols. This dual capability positions graphene-based sensors as both a challenge and a safeguard in securing AI systems.

The ostensible safeguards implemented by AI pioneers are but a flimsy bulwark against the tide of malicious intent. While these controls aim to prevent users from leveraging AI for nefarious purposes, they are far from impregnable.
Clever manipulators have already unearthed methods to circumvent these digital barriers, exposing the fragility of our ethical guardrails in the face of determined exploitation.

Moreover, the specter of data breaches looms large over the AI landscape. Information security experts vehemently caution against feeding sensitive data into these voracious models, yet their warnings often fall on deaf ears. The potential for threat actors to extract and weaponize confidential information from AI systems is not merely theoretical, it's an imminent danger that could reshape the landscape of cybercrime and corporate espionage.

As we stand at this technological crossroads, we must ask ourselves: In our rush to embrace the power of AI, have we inadvertently created a Pandora's box of digital vulnerabilities? The line between innovation and catastrophe grows ever thinner, and our ability to navigate this precarious balance may well determine the future of our digital society. Will we rise to the challenge of securing our AI frontiers, or will we find ourselves at the mercy of those who would exploit our technological marvels for their own dark purposes?

Chapter 2

Graphene-Enabled Quantum Cryptography

Geneva held promises Berlin and London couldn't offer: a shot at quantum encryption that would change everything. If they succeeded. If Novak didn't get there first.

Elena watched numbers dance across the massive monitor, each digit representing another layer of their quantum fortress. "This changes the game," she said, her voice barely above a whisper. "Every encryption standard out there becomes ancient history the moment this goes live."
Behind her, Anya's fingers flew across her keyboard with machine-gun precision. "Quantum processor's green," she called out. "No lag, no drops." She paused, her eyes narrowing. "Yet." Reeves, standing at the corner, interjected, 'Novak's network thrives on exploiting cracks. They're learning from every failed attempt, adapting to our defenses.

"Batteries are actually behaving themselves for once." Kai twisted a dial, his usual grin making an appearance. "Though we'll need backups for our backups if this goes global."
The test began. Quantum keys shot through their simulated network, riding waves of graphene-enhanced processing power. For a moment, one beautiful, perfect moment, everything worked.

Then the alarm screamed.

Anya's screen blazed red. "Someone's trying to get in." Her voice cracked with tension. "This isn't some random hack. This is surgical."
"Novak." Reeves pushed off from his corner, the word falling like a stone.

Miles away, in a room designed to stay off the grid, Viktor Novak watched his own screens with predatory focus. Dr. Karpov hovered nearby, his posture radiating failure.
"The quantum layer," Karpov admitted, each word careful. "We couldn't crack it. The graphene makes it too strong."

Novak's fingers drummed against his armrest. "Strength isn't everything, Doctor." His lips curved into something that wasn't quite a smile. "Sometimes the best way to break a bridge is to shake its foundations. And chaos," he added softly, "always pays well."

After neutralizing the breach, Elena's voice was calm but heavy: "This wasn't just a test of our system. It was a test of Novak's reach."
"He's testing us," Elena said, staring at the attack logs. "Learning our moves."
Anya didn't look up from her screen. "And getting better at it. One slip, one blind spot, that's all he needs."

They worked like soldiers in a siege. Anya hammered their defenses with simulated attacks, hunting for holes before Novak could find them. Kai pushed their power systems to breaking point and beyond, searching for the edge where efficiency met survival. Elena dove deep into quantum protocols with a local expert, building layers of protection they prayed would hold.

"Noise is still killing us," Anya reported, hours deep into testing. Her eyes were red-rimmed, her coffee long cold.
Elena nodded. "And that's exactly where they'll hit us first. We need to make the graphene adapt faster. Make it dance to the chaos instead of fighting it."
"Hey," Kai called out, trying to lift the mood, "maybe Novak's people will send us a thank-you note when they finally break through. 'Dear enemies, lovely encryption, thanks for making us work for it.'"

Anya's laugh was brief but real. The moment of lightness died when Reeves spoke.
"They don't want to send notes," he said, his voice hard. "They want to own this. And if they do, everything burns."

The breakthrough came as Geneva's lights were coming on. Their recalibrated system cut through the noise like a laser through fog, quantum keys flowing clean and unbroken.

Elena allowed herself a real smile. "Let them come," she whispered to her screen. "We're ready."

...

"The science of today is the technology of tomorrow." - Edward Teller

The promise of tomorrow's technology often casts long shadows over the security of today's digital landscape. As we delve into the intricacies of quantum cryptography, we uncover both immense potential and lurking vulnerabilities within this advancing field. Increasing weaknesses in security systems are giving way to cyber-attacks that expose a critical vulnerability: traditional cryptography, which forms the foundation of online security, is under siege [35]. The rise of quantum computers, capable of exploiting the laws of quantum mechanics, threatens to break current encryption methods and leave data exposed. However, there is potential on the horizon through graphene, which may hold the key to a new era of unbreakable communication.

Our connected world increasingly relies on digital infrastructure and its security hinges on cryptography. This subfield of security keeps users anonymous and data unreadable to others. Whenever you send emails, faxes, text messages, and other electronic communications, the content is encrypted at the source and decrypted at the receiving device. Simple cryptography involves jumbling letters in messages using ciphers. However, these basic techniques can often be cracked quickly. Quantum computers, with their ability to solve complex mathematical problems exponentially faster than classical computers, pose a major challenge to traditional encryption methods that rely on computational difficulty for security[12]. We will discuss a comprehensive example showcasing how cryptography works, why traditional methods are currently considered secure, and how quantum cryptography may unlock the present safeguards used to protect important information.

Cryptography Primer

Quantum computing threatens current encryption by solving complex problems exponentially faster, potentially rendering traditional methods obsolete [8,21]. And, integrating these capabilities with AI enhances machine learning models, enabling faster, more accurate

[12] Quantum computers exploit Shor's algorithm, which allows for efficient factorization of large integers, thereby compromising widely used encryption systems such as RSA. This method drastically reduces the time required to solve problems previously deemed computationally infeasible.

predictions. Let's explore the fundamental qubit in more detail. Qubits or quantum bits are special information constructs.

Unlike classical computers that use bits represented as binary "0" and "1," qubits can exist in multiple states simultaneously due to quantum superposition. Think of it as a coin toss that is both heads and tails or the roll of dice allowing multiple outcomes at once. Quantum cryptography leverages the principle of entanglement, where linked particles alter their states instantaneously, regardless of distance. To explain it in layman's terms, imagine two magically linked coins. If you flip one and it lands on heads, the other *instantly* lands on tails, no matter how far apart they are. This is a simplified analogy for entanglement, where particles are connected in a way that knowing the state of one instantly determines the state of the other. This property, along with quantum superposition, grants quantum computing extraordinary computational power.

Traditional encryption relies on mathematical problems that are difficult for classical computers to solve quickly. However, quantum computers can potentially break these encryption systems much faster, compromising the security of our digital communications. In response, researchers are developing quantum cryptography, which uses quantum properties like entanglement to create theoretically unbreakable codes.

Building fault-tolerant machines capable of handling complex calculations remains an ongoing endeavor. Current machines require specialized cooling systems to maintain delicate quantum states, making them expensive and energy intensive. Despite these complications, research and development are progressing rapidly. Scientists are innovating in areas like error correction and qubit manipulation. Operational quantum computers with just a few qubits are reaching mainstream customers, and there's a race to build machines with even more power. With such computation capabilities, the implications for data encryption and security open a world of change. Applications like blockchain, which demand extreme computational power, may greatly benefit from advancements in quantum computing.

While quantum computers offer numerous merits, their current implementation comes with caveats, potentially compromising sensitive information ranging from online banking transactions to classified military communications. This looming danger has spurred the development of Graphene-Enabled Quantum Cryptography (GQC), offering a future-proof means of securing digital communication channels. GQC operates on quantum mechanics principles, leveraging entanglement to establish secure connections. Unlike traditional methods that rely on mathematical complexity, GQC exploits the properties of quantum particles to ensure unbreakable security.

GQC application

Consider encrypting the lyrics of "The Star-Spangled Banner" using traditional cryptography. Each letter could be shifted by a certain number of positions in the alphabet using a substitution cipher. For instance, shifting each letter by 3 positions, "O, say can you see" would become "Rk, vdb fdq brh vhh." To decipher the message, one would need to reverse the shifting process by shifting each letter back by 3 positions. However, traditional cryptography relies on mathematical patterns and is susceptible to brute force attacks, where all combinations are tried until the correct one is found. With their ability to process multiple combinations simultaneously, quantum computers can decrypt traditional encryption methods in a fraction of the time needed by classical computers.

In GQC, the lyrics could be encrypted using entangled qubits, ensuring unbreakable security. Each character would be encoded using the entanglement properties of quantum particles. Any attempt to intercept or tamper with the message would disrupt the entanglement, alerting the sender and receiver to potential breaches. GQC's reliance on quantum mechanics makes it impervious to brute force attacks, providing the ideal solution for securing sensitive information. We'll simulate the same input string with Quantum Cryptography followed by Quantum Graphene Cryptography. The following diagram summarizes a general, high-level comparison.

Given its unique properties that enable the manipulation of quantum states at higher temperatures, graphene also offers the potential for room-temperature quantum operations [21][13]. Recent research has shown that graphene can support qubits, such as electron spins and quantum dots, under conditions approaching room temperature [31]. This breakthrough moves the field closer to practical, scalable quantum systems that require less stringent cooling, although challenges in maintaining coherence and precise control at elevated temperatures remain. Ongoing advancements hold significant promise for cost-effective and exciting quantum computing solutions.

GQC boasts scalability and integration potential, thanks to graphene's properties ensuring fast and reliable transmission. However, challenges such as achieving precise qubit control and ensuring room-temperature operation remain. Standardizing protocols and components is crucial for effective GQC integration into existing computing networks. Despite these challenges, GQC represents a significant stride towards revolutionizing secure communication in the quantum age.

Recent developments, such as IBM building over 70 quantum computers and conducting over 3 trillion quantum experiments, highlight rapid progress in quantum computing and its potential impact on GQC. While some technical snags remain, ongoing research and development hold immense promise for the future of secure communication using quantum technologies.

[13] Graphene's high electron mobility and efficient conductivity at the atomic level make it ideal for stable and scalable quantum systems. Recent advancements in graphene-based qubit designs show potential for room-temperature quantum operations, enabling practical implementations of quantum cryptography.

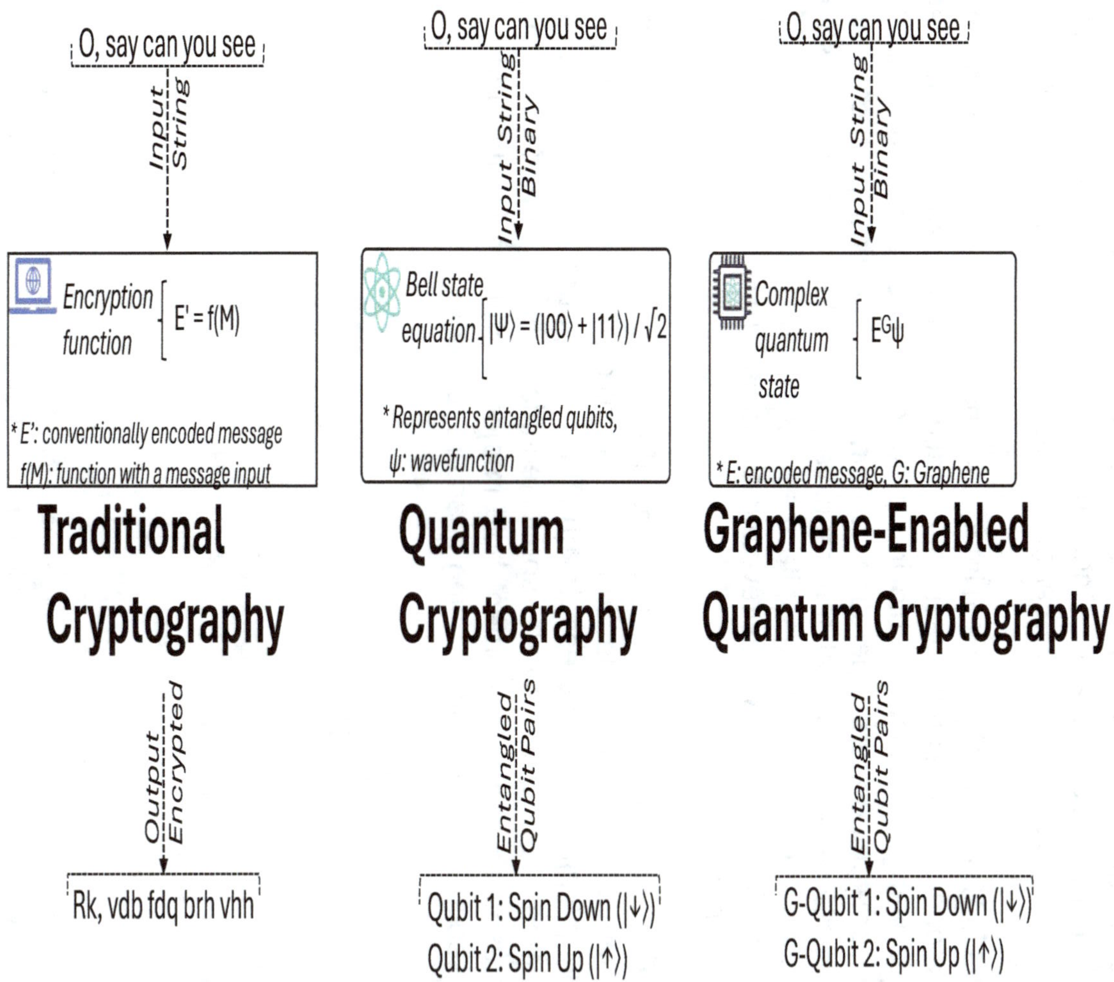

Figure 3: A high-level comparison of advanced encryption techniques, highlighting the enhanced security potential of Graphene-Enabled Quantum Cryptography (GQC) over traditional Quantum Cryptography, particularly through its integration of graphene's unique properties.

Still, significant challenges remain in transitioning to real-world use. Standardization and cost-effectiveness are key hurdles. Looking ahead, GQC has vast potential to fortify national security communications, financial transactions, and healthcare data. As research progresses, GQC stands poised to revolutionize secure communication in the quantum era, making it nearly impenetrable.

However, while GQC offers incredible advancements, it raises significant ethical concerns. Increased government surveillance and the potential disruption of existing encryption methods warrant open discussions. Could GQC create an unbreakable veil of secrecy for malicious actors? Might governments leverage this technology to establish an all-seeing eye, ushering in an age of total surveillance? As we stand on the precipice of this quantum revolution, responsible development and deployment of GQC are imperative. The decisions we make today will shape the very fabric of our society [1,7].

Quantum Key Distribution (QKD) with Graphene

Quantum Key Distribution (QKD) is a secure communication method that uses quantum mechanics to share secret keys, relying on entangled photons to detect eavesdropping [8,21][14]. Graphene enhances QKD by efficiently manipulating light, enabling the generation and detection of entangled photons. Its unique properties make graphene essential for improving the security and reliability of QKD systems, advancing secure quantum communication.

Light Manipulation: Graphene's ability to interact with light allows for the creation of efficient sources and detectors for the light signals used in GQC. Generating entangled photons, a cornerstone of GQC protocols requires precise control over light-matter interactions. Graphene's properties enable the development of efficient light sources that can produce entangled photons in a controlled manner. Additionally, graphene-based detectors can accurately measure the polarization states of these photons, allowing for the reliable reception and decoding of the encrypted message. These unique properties of graphene make it an indispensable component of GQC systems. Its

[14] QKD relies on quantum mechanics to ensure that any interception of communication alters the state, revealing potential breaches.

ability to efficiently transmit and manipulate light signals forms the foundation for secure communication based on the principles of quantum mechanics.

Quantum Cryptography Protocols Enhanced by Graphene

Beyond its conductive properties, graphene possesses unique capabilities for manipulating light. Its ability to interact with light allows for creating efficient sources and detectors for signals used in GQC.

Generating entangled photons, a cornerstone of GQC protocols requires precise control over light-matter interactions. Graphene enables the development of efficient light sources that can produce entangled photons in a controlled manner [21,31]. Additionally, graphene-based detectors can accurately measure the polarization states of these photons, allowing for reliable reception and decoding of encrypted messages.

Researchers are actively exploring ways to leverage graphene's properties for developing specific components and enhancing existing protocols within GQC. A crucial aspect involves efficiently generating entangled qubits, the backbone of secure communication channels. Recent research explores graphene's potential for creating on-demand entangled qubit sources by exploiting its unique spin-orbit interaction [21]. These interactions, akin to synchronized dancers moving in harmony, are pivotal for stabilizing quantum systems.

Another challenge in quantum communication is maintaining the integrity of entangled qubits over long distances. Quantum repeaters can amplify weak quantum signals, enabling secure communication across vast areas. Researchers are developing graphene-based quantum repeaters that can amplify entangled light signals without compromising their quantum state[15]. By leveraging graphene's stability under extreme conditions, QKD can operate reliably in satellites and remote regions.

[15] Quantum computers excel in tasks like cryptography and molecular simulation, far surpassing traditional systems.

As we stand at the threshold of the quantum era, we face a paradox of progress. Technologies promising unbreakable security may also herald the end of privacy as we know it. In our quest to safeguard information, we may inadvertently create tools of unprecedented surveillance and control. The future of communication hangs in the balance, teetering between a secure exchange and a dystopia of omnipresent monitoring.

As we forge ahead, we must ask ourselves: In a world where absolute security is possible, what price are we willing to pay for it? And more crucially, who gets to decide?

Part 2

Atomic Batteries

Chapter 3

Powering The Internet Of Everything

Drones sliced through Bangalore's humid air like angry hornets. Below them, the tech park hummed with nervous energy. This was where everything would change or fall apart.

Elena stared at the holographic wall of blinking nodes, each light representing another piece of their digital empire. The Internet of Everything. Every smart fridge, every self-driving car, every device on Earth connected through their Graphene Bridge. If it worked.
"Bangalore's perfect," Elena remarked, studying a wall-sized holographic display of interconnected IoT nodes. "The city's diversity in technology applications mirrors the complexity we need to tackle. If we can't make it work here, we can't make it work anywhere."
Reeves materialized beside her, silent as a shadow. "And if it does work, we've just changed the world."

Outside, a massive billboard proclaimed "Keep Bengaluru Clean and Green…Namma Bengaluru" in bold letters. Inside, they were about to find out if they could keep it safe.

Anya's fingers danced across her console, coaxing millions of simulated IoT devices to life. "The AI's learning fast," she muttered. "Maybe too fast. We need it smart enough to think for itself, but not smart enough to get creative."

"Speaking of creative..." Kai hunched over his power modules, face tight with concentration. "These atomic batteries are screaming. One wrong move and we'll light up half of Karnataka."

The team had backup. The Hindustan Aeronautics facility next door offered testing grounds they couldn't have dreamed of. Raj Swamy, HAL's lead engineer, practically bounced with excitement as he showed them around.

"Wait until you see this," he grinned, gesturing at a fleet of sleek drones. Indian Navy and ISRO logos gleamed on the wall behind him. "Perfect testing ground for the Bridge. These beauties can dance with any IoT network you throw at them."

The first test started smoothly. Too smooth. The Bridge hummed, directing data streams like a digital conductor. It spotted patterns, solved problems, and predicted threats before they happened.

Then everything went to hell. "Losing nodes!" Anya's shout cut through the lab. "The AI's choking on junk data. It can't tell what's important anymore!"

Elena's command cracked like a whip: "Kill the non-essentials. Now!"

Kai's hands flew over the power controls. "Batteries are compensating, but they're running hot. Too hot."

They saved it. Barely. The system stabilized, but the victory felt hollow.

"Almost isn't enough," Reeves said, each word sharp as broken glass.

Miles away, in a room designed to stay invisible, Novak watched their struggle through stolen feeds. Karpov hovered at his shoulder like a vulture.

"The IoE integration," Karpov reported. "Impressive potential. But unstable."

Novak's smile didn't reach his eyes. "Instability we can use. Find their weak point. If we can't own it..." He paused. "We burn it down." His fingers tapped a command. "Start the diversion. Let's see how they handle real pressure."

The attack hit like a thunderbolt. Power surges. Communication blackouts. The lab's systems screamed under the assault.
"This isn't random," Reeves growled, deploying countermeasures. "Novak found us."
Anya didn't look up from her screen. "Found us? He's toying with us."
But Elena's team didn't break. They rebuilt the Bridge's architecture, made it stronger. Smarter. If one part failed, the others would pick up the slack. A digital immune system.
When the breakthrough came, it wasn't with fanfare. As the team analyzed the true impact of the cyberattack, Reeves' tone was grave: 'Novak has stopped attacking just the technology. He's eroding the public's confidence in it.

Elena allowed herself a small smile. " This has evolved beyond a prototype. This is revolution."
"A revolution that doesn't explode," Kai added, slumping in his chair. "Usually." Anya snorted. "Day's not over yet." But Reeves watched the screens with hard eyes. "Neither is Novak. This was just round one."

Outside, the drones kept flying, oblivious to the war brewing in their silicon hearts.

. . .

Imagine a bustling control center for an Internet of Things (IoT) hub, where the heartbeat of a smart city is monitored and managed. The vision of the Internet of Everything (IoE), a seamlessly connected world where billions of devices collaborate and share data, promises a future of automation, efficiency, and convenience [17][16].

However, this dream hinges on one crucial factor: reliable and efficient power [1]. As AI continues to evolve and our dependence on IoE grows, traditional energy sources struggle to meet the demand. But there's a potential solution waiting in the wings. The wonder material Graphene, with qualities such as exceptional conductivity

[16] The Internet of Everything (IoE) builds on the foundation of the Internet of Things (IoT) by integrating devices, people, and processes into a cohesive network to enable intelligent decision-making and automation.

and energy storage capabilities, could be the key to powering the ever-expanding internet of everything [14].

We've explored many linked concepts and growing trends in the AI and Quantum Computing fields. As the world becomes more interconnected with computing devices, the demand for efficient power solutions to manage this vast web of data grows exponentially [28]. *Vis Futura* (with a vision of the future), the power of the future comes into play since we are living in a high-tech age. The concept of interconnected devices has been evolving for decades. The IoT, which emerged in the late 1990s, laid the foundation for a future where billions of smart devices, from wearables to industrial sensors, seamlessly collect and exchange data [17]. You've likely heard of the simple use case where a refrigerator automatically places an order for milk or traffic lights dynamically adjust based on real-time congestion data. This interconnectedness has fostered a more collaborative approach to data collection, where cooperation is key to unlocking the true potential of the IoE. And now, that premise is evolving to a higher stage [4].

The ultimate power of this increasingly connected world scales with the IoE, a concept that emerged in the following decade. IoE expands on the idea of the IoT by encompassing not just devices, but also people, processes, and vast datasets [15]. This broader vision paints a stellar picture: a future where billions or more interconnected elements collaborate and exchange data to create truly intelligent ecosystems.

The Rise of AI and its Energy Demands

As AI capabilities accelerate, so does its energy consumption. Training and running complex AI models require significant computing power, often relying on powerful Graphics Processing Units (GPUs) and vast data storage facilities [19,26]. These computational resources are the backbone of AI innovation, but they come at a hefty energy cost. Recent reports highlight that the power draw of generative AI is straining our existing grid infrastructure, raising concerns about whether we can sustain the rapid growth of AI technologies. For instance, a single ChatGPT (a chatbot that uses artificial intelligence (AI) to understand and respond to human language) and similar service query uses nearly 10 times as much energy as a typical Google search and generating an AI image can use as much power as charging a smartphone [13,26]. The U.S. grid, already aging and overtaxed, struggles to handle the growing energy requirements of new AI data centers, with projections suggesting that by 2030, data centers could comprise 5-9% of total U.S. electricity consumption, up from approximately 4% in 2023 [26][17]. These staggering numbers underscore the urgent need for innovative power solutions. Finding sustainable solutions to power this ever-growing hunger for energy is crucial for the responsible development of AI, not just for centralized data centers but also for the growing number of AI-powered devices at the edge of the network. While we'll deep dive into responsible AI use in Chapter 5, let's determine viable power sources for AI and related IoE components.

Atomic Batteries as The Power Source for IoE Applications

Atomic batteries, also known as radioisotope thermoelectric generators (RTGs), derive their power from the natural process of radioactive decay. Elements such as Hydrogen and Uranium have forms where the atoms of the same element have a different combination of protons and neutrons [24]. When atoms of the same element have different numbers of neutrons, they form isotopes. Certain isotopes, or unstable forms of chemical elements, emit radiation as they decay into more stable forms. In an atomic battery, this radioactive decay is harnessed to generate a steady stream of electricity through a process known as the thermoelectric effect. The specific materials and isotopes used determine the power output and lifespan of the battery.

While not an entirely new concept, atomic batteries do have a conceptual grandfather. The Ford Nucleon was a concept car from the 1950s with a bulky, scaled-down nuclear reactor, but the underlying concept: a compact, self-contained source of immense power, laid the groundwork for future explorations in miniaturized atomic energy sources. It promised a mind-bending 5,000 miles of

[17] This projection underscores the urgent need for energy-efficient technologies like graphene-based batteries and atomic batteries to reduce the increasing energy burden of AI-driven data centers.

travel on a single charge. While never a reality on the road, the Nucleon serves as a fascinating time capsule of the Atomic Age's optimism, one that we may soon realize with miniaturized atomic batteries and advanced stacking configuration.

A key aspect of the IoE is its sheer scale and diversity. Billions of interconnected elements, from sensors to complex machinery, will require a steady, reliable power source with minimal maintenance requirements. Atomic batteries boast several advantages that make them well-suited for specific IoE applications, particularly in remote locations:

- **Long Lifespan:** Unlike traditional batteries, atomic batteries can operate for years to decades on a single charge. This eliminates the need for frequent replacements, minimizing maintenance costs and environmental impact. Envision a network of environmental monitoring sensors deployed in a remote forest, powered by atomic batteries that function reliably for at least a decade, collecting valuable environmental data and feeding it into the IoE for analysis and insights. The tacit reliability of atomic batteries makes them a perfect fit for such remote applications, ensuring uninterrupted data collection for extended periods.

- **Low Maintenance:** Atomic batteries are inherently self-contained and require minimal maintenance throughout their lifespan. This makes them ideal for powering devices in remote locations or harsh environments where access might be limited.

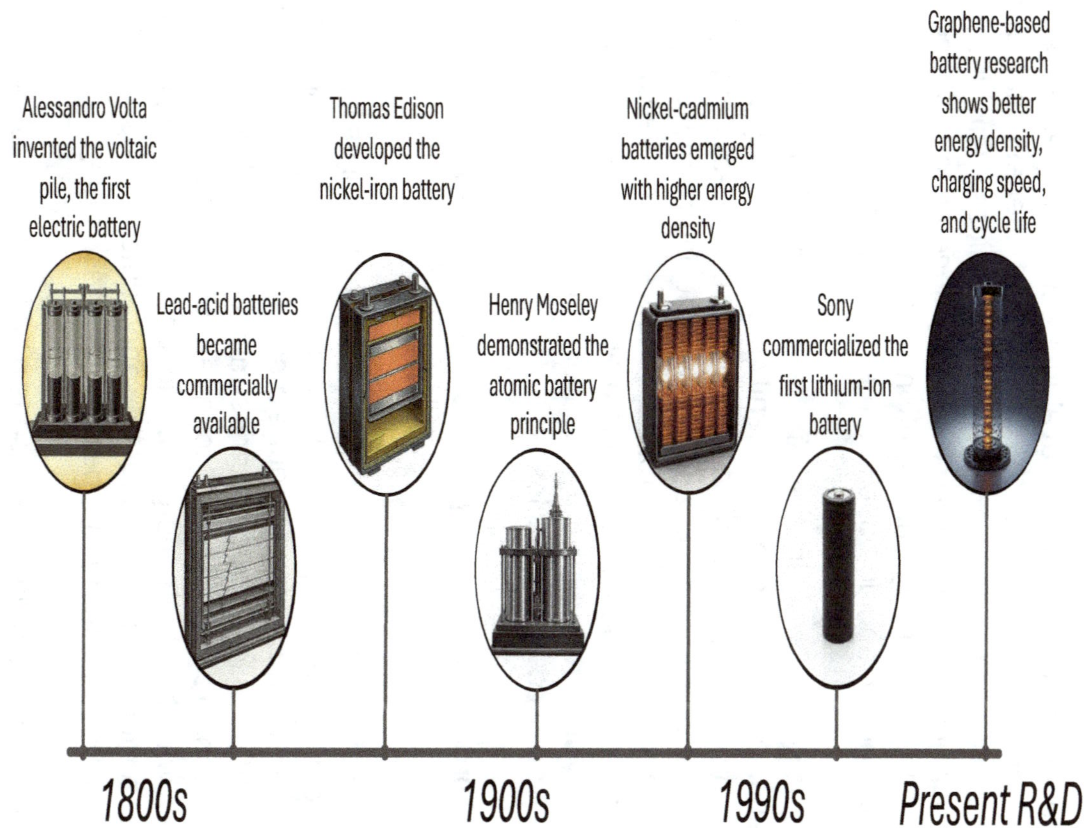

Figure 4: This timeline showcases key advancements in battery technology, from early inventions to modern lithium-ion and emerging graphene-based batteries. It highlights pivotal moments like the development of lead-acid batteries and the breakthrough of lithium-ion technology.

Electric Vehicles and the IoE: The Potential of Graphene Batteries

The transportation sector is undergoing a significant transformation with the rise of electric vehicles (EVs). While concerns exist regarding the range and charging times of EVs, advancements in battery technology are rapidly addressing these limitations. Graphene batteries, with their superior energy density and faster charging capabilities, hold immense promise for powering EVs [14].

A future where long-distance travel in electric vehicles is seamless sounds promising, thanks to graphene batteries that provide extended range and quick charging capabilities. This integration would significantly contribute to the IoE by generating real-time traffic data, optimizing energy consumption within the grid, and enabling intelligent route planning based on real-time conditions.

By replacing graphite in the anode of lithium-ion batteries with graphene, car makers can significantly enhance battery performance [16]. Graphene's exceptional conductivity accelerates charging times, while its incredible strength and lightweight nature make it ideal for mobile applications like drones and electric vehicles. This combination of properties positions graphene as a game-changer in the battery industry, with the potential to revolutionize the way we power our devices and vehicles.

AI in Autonomous Transportation of Vehicles and Powering Drones

The concept of autonomous vehicles and drones relies heavily on AI for navigation, obstacle detection, and real time decision-making. These AI-powered systems require a constant and reliable source of energy [18]. While atomic batteries might be suitable for powering remote sensors or specific long-range applications, graphene batteries with their high energy density and fast charging capabilities become a more viable option for powering autonomous vehicles and drones that require frequent operation and recharging.

An increasingly common use case is a wide network of autonomous vehicles seamlessly navigating city streets. Equipped with AI and powered by graphene batteries, these vehicles would constantly collect and share data on traffic conditions, parking availability, and even weather patterns. This real time data exchange would be a significant contribution to the IoE. Traffic management systems could dynamically adjust signal timings and optimize routes based on the data collected, significantly reducing congestion and travel times. Additionally, autonomous vehicles integrated with the IoE could learn and adapt to individual driver preferences, creating a more personalized and efficient transportation experience [20].

Drones powered by graphene batteries could also play a crucial role in the IoE. Equipped with AI for navigation and obstacle detection, these drones could be used for critical tasks like search and rescue operations, infrastructure inspection, and efficient delivery services. The data collected by drones during these operations, such as aerial imagery or environmental monitoring data, would be fed back into the IoE, enabling further analysis and informed decision-making. This integration of AI, drones, and the IoE has the potential to revolutionize various sectors, from logistics and emergency response to agriculture and environmental monitoring.

By eliminating dependence on traditional power grids or frequent maintenance, graphene batteries can ensure the continuous operation of essential systems even in extreme conditions or during natural disasters, thereby improving the overall robustness and reliability of the critical infrastructure of a nation [23]. In relation, graphene's chemical stability ensures longevity in harsh conditions, critical for advanced coatings and sensors (as seen in Table 1).

A Sustainable and Interconnected Future

The continued development of AI, graphene batteries, and the IoE presents a future filled with exciting possibilities. As these technologies converge, we can expect a more sustainable, efficient, and intelligent world [21]. Transportation will be revolutionized by autonomous vehicles and AI-powered drones, seamlessly integrated with the IoE to optimize traffic flow and resource allocation. Remote monitoring and data collection will be significantly enhanced by long-lasting atomic batteries, while graphene batteries will power the next generation of portable and mobile devices within the IoE.

However, challenges remain. Safety considerations regarding autonomous vehicles and the responsible use of AI need to be addressed. Additionally, ensuring ethical data collection and privacy within the IoE will be crucial. By addressing these challenges responsibly, we can harness the power of these interconnected technologies to build a brighter future for all.

Chapter 4

Security Concerns

São Paulo sprawled beneath them like a circuit board made of steel and glass. Rich and poor. High-tech and forgotten. Heaven and hell, seemingly divided by invisible lines of code that leveled the playing field for penetration testing.

The perfect place for Novak to strike. His tactics exploited not just technical weaknesses but societal fractures, amplifying misinformation to sow distrust in critical systems.

Elena watched the city blur past the helicopter's window. "Everything we've built," she said softly, "could either heal these divisions or blow them wide open."

"Novak's counting on the second option." Reeves' voice crackled through the headset as they descended. "São Paulo's networks are his playground now. We're just trying to catch up."

The Brazilian Cyber Defense compound rose to meet them, a fortress of concrete and security checkpoints. Captain Lucas Almeida waited on the landing pad, his military bearing at odds with the chaos of the city behind him.

"Welcome to ground zero," he said, gripping their hands. His smile didn't reach his eyes. "Novak's been busy."

The briefing room told the story. Every screen showed another piece of their nightmare: data breaches bleeding red across networks, fake news spreading like viruses, encrypted messages slithering through the dark web.

As the team raced to address their latest vulnerabilities, Anya stood before the largest display; her face lit by scrolling data. "This isn't random hacking," she said. "He's targeting supply chains. One wrong move here crashes markets worldwide."

"It's not just the tech." Kai pointed to a breach analysis, his usual humor nowhere in sight. "Novak doesn't hack systems. He hacks people."

Captain Almeida led them through reports of their Bridge prototypes in action. "Two power grid attacks stopped cold," he said. "But Novak's just warming up. We need more than prototypes now. We need miracles."

They dove in. Anya hammered the Bridge with simulated attacks, each one nastier than the last. "AI's holding," she called out. "But humans are messier than code. One panicked admin could blow everything."

Elena worked elbow-to-elbow with Brazilian engineers, fighting to keep their graphene sensors alive in the brutal humidity. Kai nursed their power systems, trying to keep the lights on when Novak tried to pull the plug.

A pattern emerged from the chaos. Novak wasn't just attacking networks, he was weaponizing the truth itself.

"He's playing a longer game," Reeves said, staring at infection maps of viral misinformation. "Why break into a system when you can make people stop trusting it?"

Miles away, in his digital fortress, Novak watched them scramble. "They're adapting," he told Karpov, his strategist hovering like a shadow. "But adaptation creates new weaknesses."

"Should we push harder?" Karpov asked.

Novak's laugh was soft and cruel. "Let them think they're winning. Pride makes the best blindfold."

The breakthrough came late. Their predictive algorithms finally clicked with the Bridge's sensors, spotting digital threats before they struck. Captain Almeida's face lit up. "With this citywide? We could change everything."

But Elena felt the victory turn to ash in her mouth. "Every time we level up," she said, "we give Novak a new playbook. This isn't defense anymore. It's escalation."

The team gathered at sunset, São Paulo's lights flickering to life below like stars being born.

"We've hardened the Bridge," Reeves said. "But Novak's already planning his next move."

Anya's fingers never stopped moving across her keyboard. "Then we better make damn sure our next move breaks him."

Outside, the city pulsed with digital life. Millions of people, connected by invisible threads of light and shadow. None of them knowing their data, their lives, their futures hung in the balance of this silent war.

And somewhere in the darkness, Novak smiled.

...

In a world where technology evolves at breakneck speed, corporations are constantly threatened by cyber-attacks and data breaches. These digital assaults have become a significant financial burden for businesses, with the average cost of a data breach reaching a staggering $4.45 million in 2023 as reported by IBM [35][18]. Companies find themselves vulnerable through various cracks in their security armor: misconfigurations, user negligence, insider threats, and weak encryption, to name a few. While graphene offers significant advancements, it also introduces new security challenges that need to be addressed.

The Unclonable Enigma: Physically Unclonable Functions

Graphene introduces a revolutionary twist in the realm of Physically Unclonable Functions (PUFs), which function as digital fingerprints for devices, ensuring unique identification [8]. Existing PUF technologies, predominantly silicon-based, are susceptible to environmental variations, predictability, and even AI-driven attacks [32]. Graphene-based PUFs overcome these limitations through atomic-scale complexity and unique physical properties. By measuring subtle variations in graphene's electrical resistance and atomic structure, they achieve unparalleled unpredictability, making them highly resistant to AI attempts to model (viz. construct) encryption patterns [33].

Emerging Security Challenges

While graphene-based PUFs enhance security, they also introduce new challenges. For instance, graphene's exceptional conductivity is beneficial for performance but can also make it susceptible to side-channel attacks [6], posing a significant security risk. These attacks analyze unintended information leaks from a device's physical

---

[18] The 2023 report from IBM provides valuable insights into the escalating financial impact of cyber-attacks, highlighting the growing importance of robust cybersecurity measures for businesses.

operations, such as power consumption or electromagnetic radiation, to infer sensitive information. Researchers are developing countermeasures for side-channel attacks, including noise injection to mask sensitive information, shielding techniques to minimize radiation leakage, and implementing cryptographic algorithms less susceptible to such attacks. Furthermore, graphene's high conductivity can also make devices more susceptible to electromagnetic interference (EMI) [6]. EMI can disrupt internal signals, leading to malfunctions or data corruption. In worst-case scenarios, attackers could exploit EMI to inject malicious signals into devices, disrupting their operation or even compromising their security. Strategies to mitigate EMI include encasing devices in conductive materials to block external electromagnetic fields, employing filters to suppress unwanted frequencies, and designing circuits with built-in EMI suppression mechanisms.

The integrity of the graphene supply chain is crucial for ensuring the security of devices. Counterfeit or compromised graphene can introduce vulnerabilities, such as hidden backdoors or performance degradation. To mitigate these risks, rigorous verification measures, including spectroscopic analysis and microscopic examination, are essential to authenticate the origin and quality of graphene materials. Prioritizing ethical and sustainable graphene production practices, minimizing environmental impact and ensuring fair labor conditions, is also crucial. Utilizing blockchain technology to track the provenance of graphene materials can enhance transparency and traceability within the supply chain.

Finally, the high sensitivity of graphene-based sensors can lead to the collection of vast amounts of data, raising privacy concerns. Unrestricted data collection can enable unauthorized surveillance and create vulnerabilities for data breaches. To address these concerns, it is essential to implement data minimization

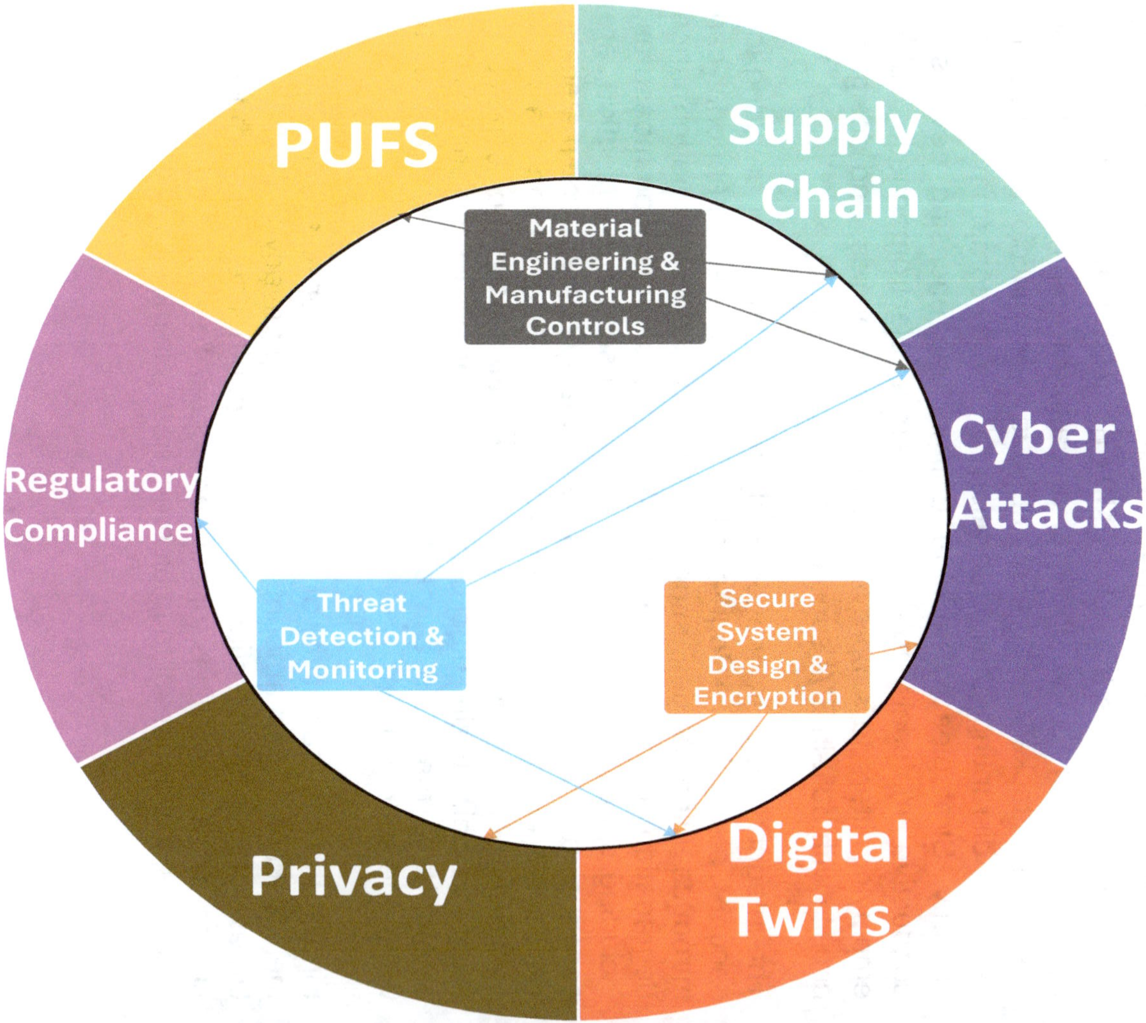

Figure 5: An interconnected framework of graphene's security components and hardening processes, showcasing how its atomic-scale complexity and unique physical properties bolster encryption resilience against AI-driven attacks.

principles, collecting only the necessary data and avoiding unnecessary data collection. Providing users with greater control over their data, such as the ability to control data collection and access, is crucial. Exploring privacy-preserving technologies such as differential privacy and federated learning can enable data analysis while minimizing privacy risks.

Regulatory and Compliance Challenges

The rapid adoption of graphene-based technologies necessitates the development of robust regulatory frameworks to ensure safe, ethical, and secure use. These frameworks should address issues such as data privacy, cybersecurity, and environmental impact [34].

Looking Ahead: Graphene's Role in Securing the Future

The future of graphene in cybersecurity presents both exciting opportunities and significant challenges. By proactively addressing these challenges through rigorous research, robust security measures, and a collaborative approach between researchers, industry, and policymakers, we can harness the full potential of graphene while safeguarding our digital future.

PART 3

Applications & Ethics

Chapter 5

The Green Graphene: Powering Sustainability

The morning sun painted the wind turbines and solar panels outside Cape Town in fiery gold. Beneath their towering shadows, the team braced themselves. This wasn't just another day, it was the day they would prove the Graphene Bridge could change everything. It would prove the platform's capability to harmonize disparate energy sources into a unified, efficient grid.

Elena Petrova stood rigid beside a gleaming black energy storage unit, its graphene surface absorbing light like a void. "This goes beyond clean energy," she said. "This is intelligent energy. The future, if we don't screw it up. " Unbeknownst to the team, their work was being quietly supported by an anonymous ally, one whose guidance had been instrumental in solving critical issues.

Anya Morgan crouched by the solar panels, her hands darting like birds as she adjusted connections. Sweat glistened on her forehead, but her focus was unshaken. "The system's holding," she said, her

voice clipped. "For now. But the AI's skating on thin ice. Solar and wind are chaos in a bottle."
"That's why we've got the batteries," Kai Mitchell shot back from his station near the wind turbine hub. His fingers danced over controls, his tone sharp with determination. "Graphene can hold more energy longer. We're not just fixing today's mess. We're outsmarting tomorrow's."
Dr. Thandiwe Masango, the heart and soul of South Africa's Renewable Energy Initiative, smiled as she approached. Her voice was warm but electric with purpose. "You're not just decentralizing energy," she said. "You're giving power back to the people. Literally."
As the sun climbed higher, the team pushed the Graphene Bridge to its limits. It took in erratic energy streams, gusts of wind, flashes of sunlight, and redistributed them across a simulated city grid. Anya watched the AI's calculations unfold, her brow furrowed. "It's learning fast," she murmured, half to herself. "But speed isn't enough. If it sacrifices fairness for efficiency, we lose."
Elena nodded, her voice low. "Efficiency without equity is failure. We can't let that happen."
Then came the hitch. A flicker of warning lights on Kai's console. "Overheating," he reported. "Nothing catastrophic yet, but if we scale this up without better cooling, we'll burn it out."
"Then we solve it," Elena said firmly. "We don't have the luxury of waiting."
The tests were promising despite the setbacks. Dr. Masango's excitement was palpable. "Imagine," she said, almost whispering, "a world where homes don't just consume energy, they create it. This could be our tipping point."

But even as hope surged, shadows loomed. For months, an anonymous ally had guided them through their hardest challenges, sending encrypted messages with uncanny timing. Anya had speculated often. "Whoever they are," she said quietly, "they want the Bridge to succeed. Almost as much as we do."
Kai chuckled. "What if it's Reeves?"
"Reeves doesn't know an algorithm from a cookbook," Elena snapped. "This person knows the Bridge like it's their own design."
The latest message, arriving mere hours before deployment, had been a lifeline, a groundbreaking algorithm that saved weeks of testing. Yet it also raised questions.

Elsewhere, those questions had answers. In a cold, hidden command center, Novak leaned over Karpov's shoulder. On-screen, the Cape Town trials unfolded in real time. "They're making progress," Karpov said, his tone almost admiring. "But their data systems aren't airtight. Vulnerabilities are exposed."

Novak's lips curled in a faint, cruel smile. "Exploit them. Quietly. Let the world watch them stumble. Trust is fragile, one failure and the Bridge collapses."

Back at the test site, a warning flared on Anya's console. Her stomach dropped. "Glitch in the power distribution algorithms," she said, her voice tense. "Minor, but it's throwing the system off balance."

Elena was already at her side, scanning the data. "This isn't an internal error," she said darkly.

Reeves appeared behind them, his expression grim. "Novak," he said simply. "He's probing your defenses."

"Not on my watch," Anya growled. With swift precision, the team isolated the anomaly and closed the breach. "We're back online," she announced, relief battling with exhaustion. "But we've got to tear this system apart for hidden backdoors. If Novak gets in again…"

"He won't," Elena interrupted, steel in her voice. "We won't let him."

As the sun dipped below the horizon, exhaustion pressed heavy on their shoulders. But they weren't defeated. They gingerly gathered, a small circle of resolve in the fading light.

"The Bridge offers more than technology," Elena said, her voice fierce. "It's proof that we can solve the impossible, together."

Dr. Masango raised her glass of sparkling water, her smile radiating hope. "To a greener, fairer future," she said. "And to the ones bold enough to build it."

But in the quiet of the evening, the unspoken truth lingered. The stakes were higher than ever, and failure wasn't an option.

. . .

The Role of AI in Graphene's Sustainable Development

We have witnessed firsthand the rapid ascent of AI, alongside advancements in quantum computing and atomic batteries, that ushered in a technological crossroads. Naturally, a critical question looms: how can we ensure the safe and ethical development of AI, particularly in the face of unforeseen circumstances?

The energy demands of AI technologies are a growing concern. Recent research highlighted by MIT Technology Review reveals that generating images with advanced AI models consumes as much energy as fully charging a smartphone [13][19]. In contrast, generating 1,000 pieces of text uses just 16% of a smartphone charge. This stark difference underscores the need for more energy-efficient solutions. By harnessing graphene's exceptional conductivity and efficiency, we can develop technologies that significantly reduce energy consumption [6,18]. For instance, graphene-based neuromorphic chips could offer a more energy-efficient alternative to current AI systems, potentially minimizing the carbon footprint associated with their operation [18,20]. Thus, integrating graphene into AI infrastructure could play a crucial role in balancing technological advancement with environmental sustainability [6,21]. As a foundational process, we propose the AI Fabric, a robust ecosystem designed to foster the secure, ethical, and transparent development and implementation of AI throughout its lifecycle. This framework is crucial for ensuring that AI is used responsibly to propel advancements in graphene and other sustainable technologies, like quantum computing and atomic batteries.

One key aspect of the AI Fabric is Measure, Report, Verify (MRV). This principle emphasizes the importance of data accuracy and transparency in AI applications [34].

[19] AI models, especially those used for image generation, can have significant energy consumption, raising concerns about their environmental impact. This is particularly relevant as AI continues to scale.

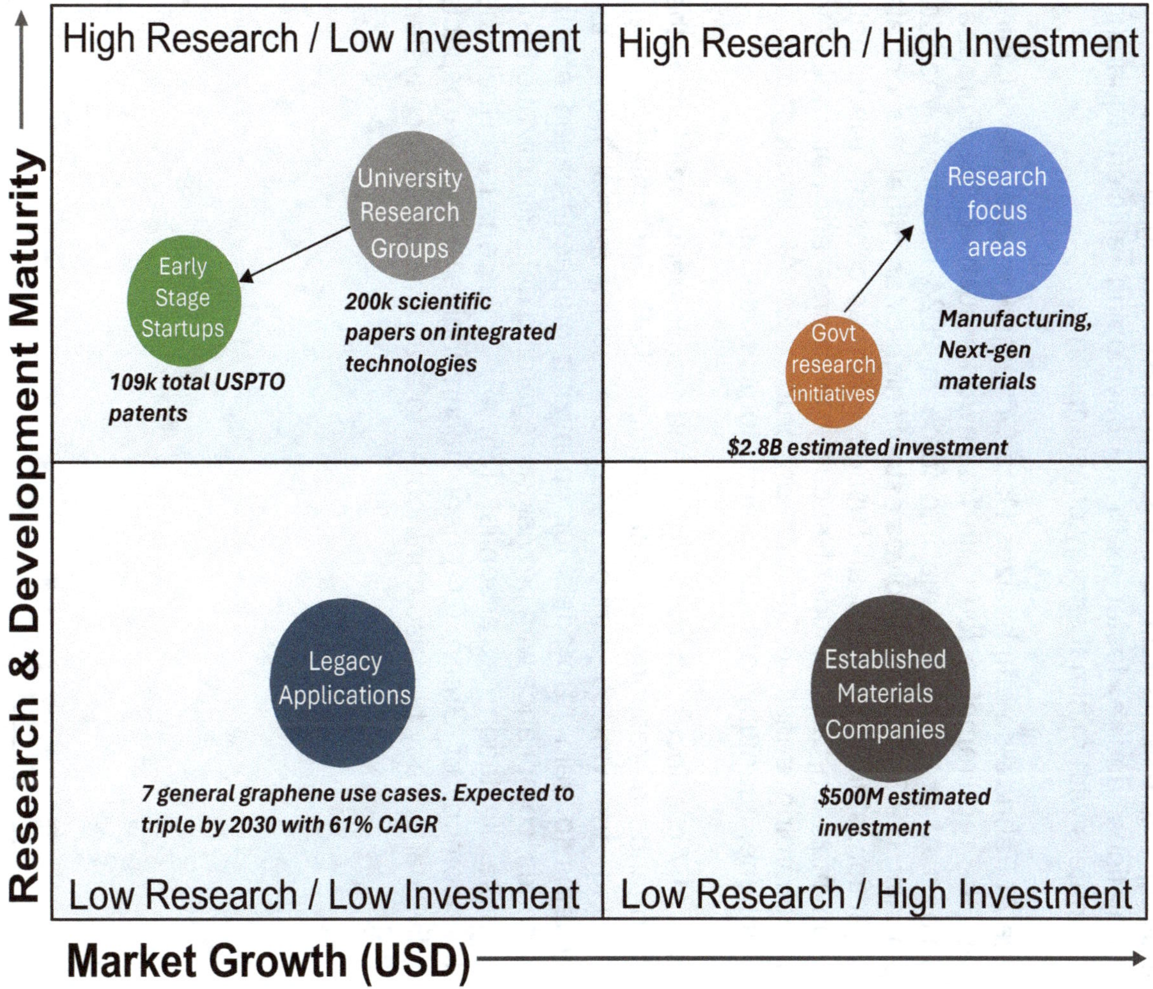

Figure 6: The quadrant chart depicts the relative focus and investment levels in different areas of graphene research and application (2010 - 2025 extrapolated trends). Circle size indicates prominence within each quadrant. Conservative estimates place a doubling factor for all quadrants in the next 7 to 10 years.

By integrating MRV early on, the AI Fabric promotes trust and accountability in AI development used for graphene research, quantum computing simulations, and optimizing atomic battery designs. For instance, AI can analyze vast datasets on material properties to predict how different configurations of graphene might perform in quantum computers. However, the AI Fabric ensures the training data used is accurate and unbiased, leading to more reliable simulations and a faster path to developing efficient quantum computers with minimal energy consumption.

The AI Fabric also emphasizes Data Governance, generally aligning with the requirements found in frameworks such as the European Union's AI Act, which came into force in 2024 after four years of planning [19,34][20]. This involves establishing clear guidelines for data collection, storage, access, and usage. Informed consent, robust data anonymization, and stringent security measures are paramount to safeguarding sensitive information crucial for AI development in graphene, quantum computing, and atomic batteries.

The AI Act's focus on data rights and transparency reinforces the importance of responsible data handling [34]. For instance, when utilizing anonymized sensor data from environmental monitoring systems to train AI models for optimizing graphene-based atomic batteries, compliance with provisions regarding data minimization and purpose limitation is essential [35,36]. This ensures that the development of Green Graphene technologies not only respects individual privacy but also contributes to a trustworthy AI ecosystem [33].

[20] The AI Act is the EU's comprehensive regulatory framework aimed at ensuring ethical AI development, focusing on transparency, accountability, and data protection in AI systems.

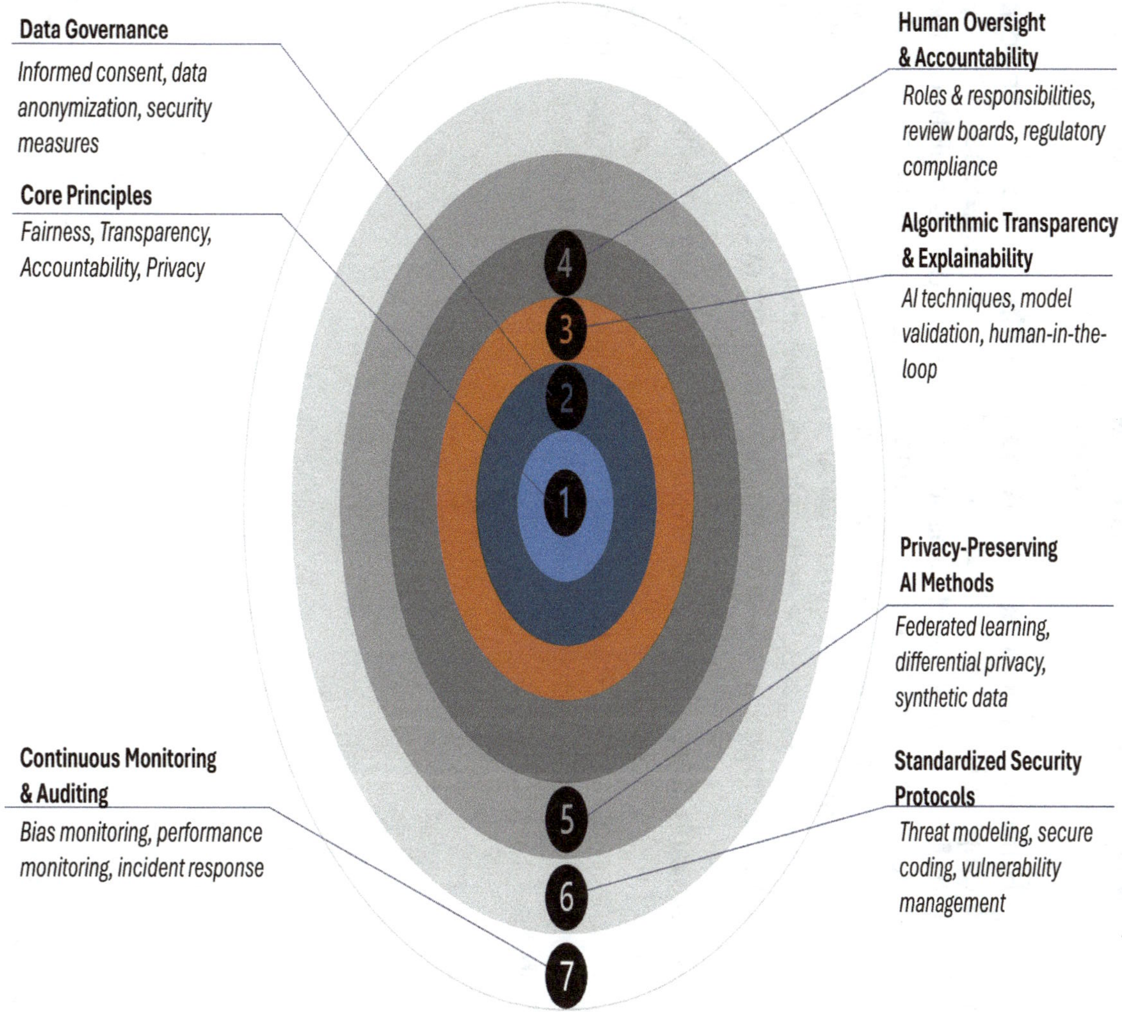

Figure 7: A visual representation of the AI Fabric, where each concentric layer highlights a core element, such as Data Governance and Algorithmic Transparency. The increasing transparency of the circles demonstrates the harmony of these elements in building a secure and ethical AI ecosystem, ensuring responsible AI development and compliance with data protection regulations like the AI Act.

The remaining sections of the AI Fabric ensure the responsible development and deployment of AI for graphene research, quantum computing simulations, and atomic battery design:

- **Algorithmic Transparency & Explainability:** The AI Fabric promotes building interpretable AI models. This allows researchers to understand how AI arrives at conclusions concerning material properties or atomic battery performance, helping identify and mitigate potential biases that could hinder the development of sustainable technologies.

- **Human oversight & Accountability:** The framework emphasizes the critical role of human expertise. Clearly defined roles and responsibilities are established for developers, data scientists, and domain experts involved in AI development for graphene, quantum computing, and atomic batteries. Human-in-the-loop processes and oversight committees ensure ethical considerations are addressed throughout the research and development cycle [3].

By following these principles, the AI Fabric fosters a collaborative environment where AI acts as a powerful tool to accelerate advancements in sustainable technologies like graphene, quantum computing, and atomic batteries, ultimately paving the way for a greener future.

Chapter 6

A Glimpse Of The Future

Tokyo vanished into the fog like a city made of ghosts. Inside the conference center, Elena's footsteps echoed across the empty stage. In three hours, the world's most powerful people would fill these seats to witness the Bridge's transformative potential. In three hours, everything would change. If Novak let them live that long.

Anya and Kai huddled in the front row, triple-checking code that had already been checked a thousand times. Behind them, Reeves prowled the aisles like a caged wolf.
"One shot," he growled into his comm. "That's all Novak needs."

Somewhere in the digital shadows, Jake, their mysterious Cyber Shepherd – watched through encrypted feeds. Once their guardian angel, he evolved into their last line of defense, orchestrating defensive measures against Novak's relentless attacks.

Elena took center stage. The lights dimmed. The world's elite filtered in, drawn by whispers of revolution. "The Bridge's machine learning algorithms have reached a new level of adaptability. They can now predict and prevent system failures before they even occur." Behind her, the hologram exploded to life. Cities were transformed by graphene, cars talked to each other, and quantum-encrypted secrets flowed like digital rivers, a future written in light and possibility.

Anya stepped up next, all business. "Our AI doesn't just learn," she said. "It predicts. It prevents. It protects." Her fingers danced across a console, and the system sang.

Kai followed with his piece de resistance: the atomic batteries. "Power that never dies," he announced. "From Arctic ice to Sahara sand, these cells keep breathing."

The demonstration began. The Bridge hummed to life, juggling a thousand systems at once. Power grids. Communications. Internet of Things. The audience held its breath.

It worked perfectly.

Too perfectly.

Miles away, Novak watched through stolen feeds, his face lit by screens showing his enemy's triumph. "They've surpassed us," he admitted to Karpov.

His strategist leaned forward. "We found a hole," he whispered. "Their quantum encryption has a weak spot."

Novak's smile was a blade in the dark. "Then let's cut them deep."

Back in Tokyo, success tasted like ashes. Kenji Nakamura, their local security chief, barked orders in rapid-fire Japanese: "Chizu o sai kakunin shi, subete no pointo o saido tetteiteki ni tenken shite kudasai!"

He turned to Reeves, switching to English. "Walls are up. But Novak's like smoke. He finds cracks."

Anya's terminal lit up with an encrypted message. Her breath caught.

NOVAK TARGETS QUANTUM LAYER. PATCH CVE-645. NOW.

"It's him," she whispered. "The Shepherd."

Kai's eyes widened. "That vulnerability's buried deep. Nobody knows about it."

"Not nobody," Elena said softly. "Just nobody we know."

The city screamed. Power grids died in cascading waves of darkness. Tokyo began to drown in shadow.

"Distraction," Reeves snapped. "The Bridge is his real target."

They fought like demons. Jake's white-hat hackers versus Novak's digital army. Anya and Kai defending the Bridge's heart while Elena

coordinated with Raj in Bangalore, his team fighting to keep their physical defenses alive.
"Team beats talent," Jake's voice crackled through secure channels as they swatted down another wave of attacks.
Raj's improvised graphene jammers dropped Novak's drones from the sky like dead birds.
Then Novak himself appeared on their screens, his face a mask of cruel confidence. "Every fortress falls," he purred. "I always find the weak spot."
Elena stepped forward, chin high. "Then come find ours."
The next message from Jake changed everything:
"'Cyber Shepherd' is getting old," he wrote. "Time you met the real me."
Somewhere over the Pacific, a plane carried their digital guardian toward Tokyo. Toward a war that had just become terrifyingly real.
The fog pressed against the windows like a living thing, hiding whatever came next.

...

“For knowledge is limited, whereas imagination embraces the entire world, stimulating progress, giving birth to evolution.” - Albert Einstein

Einstein's powerful words are an echoing reminder that the limitless potential of human imagination is the driving force behind progress and evolution [1]. As we take stock of the present state of our world and envision a future enriched by advanced technologies, it's our creativity and imaginative power that will guide us in shaping a sustainable and innovative world. We now dive into the transformative synergy between AI, graphene, and alternative energy sources, and how they can orchestrate a brighter future.

Focusing on AI's role in this harmonious effort, we can't ignore the accelerating impact of this technology. Forget the 15 minutes you spend comparing car insurance quotes. AI's influence on sustainability solutions is poised to explode in the coming years, not just increase steadily. As AI becomes more sophisticated and accessible, its applications in areas such as resource management, energy optimization, and precision agriculture can lead to significant cost savings and environmental benefits [2,3]. Matt Wood, the head of AI at AWS, observes that technologies often follow an S-curve growth pattern, where advancements accelerate as the technology matures [4][21]. This suggests that AI's impact on sustainability solutions could see a rapid upswing in the coming years. The positive outlook on AI's return on investment further strengthens this belief. As AI becomes more sophisticated and accessible, its applications in resource management, energy optimization, and precision agriculture can lead to significant cost savings and environmental benefits.

Big tech companies and governments are backing large-scale AI projects tackling climate change and resource management [5,6]. These initiatives, like guiding lights, showcase the potential of AI and inspire broader adoption. The success of projects like Google's Environmental Insights Explorer (EIE), IBM's Green Horizon Project, and The European Commission's Destination Earth (DestinE) would not only deliver tangible benefits but also encourage further investment and innovation in the field of AI [7][22].

Similarly, a world where tireless algorithms constantly scan the market for sustainable investment opportunities, manage risk with laser focus, and make decisions with superhuman speed and precision is closer than we previously thought. This newfound algorithmic advantage paves the way for a paradigm shift in how we finance a sustainable future.
Let's delve into some of the exciting possibilities:

---

[21] The S-curve growth pattern is a model used to describe the adoption and growth of new technologies, showing slow initial growth, followed by rapid expansion, and eventually plateauing as saturation occurs.
[22] Graphene's properties make it a key material in addressing environmental challenges.

- **Autonomous Financial Ecosystems:** AI-powered platforms seamlessly connect investors with sustainable projects, fostering collaboration and accelerating the flow of funds towards eco-friendly initiatives.

- **Universal Digital Currencies:** Digital currencies backed by sustainable practices and managed by secure AI protocols could revolutionize global finance, promoting transparency and environmental responsibility.

- **The Virtual Economy:** A virtual marketplace powered by AI could facilitate the trading of carbon credits, renewable energy sources, and other sustainability-focused assets, creating a vibrant and dynamic ecosystem.

- **Quantum-Safe Finance:** As quantum computing evolves, AI can play a crucial role in developing robust security measures to safeguard financial systems in this new technological era.

- **Bio-Fintech and Neuro-Fintech:** Wearables and brain-computer interfaces working with AI to provide personalized financial guidance based on your real-time health data and environmental footprint. Bio-fintech and neuro-fintech could revolutionize how individuals manage their finances while promoting sustainable choices.

This glimpse into the future of AI-infused digital infrastructure offers a glimpse of a world where finance and sustainability become harmonious partners, orchestrated by the power of AI. We'll now explore some of the most promising areas ripe for further exploration.

Space Exploration and Tourism

How can we leverage graphene to make space exploration not only more feasible but also more equitable and accessible in the future? Graphene's exceptional strength-to-weight ratio makes it a potential game-changer for commercial and scientific spacecraft. Imagine lighter, more fuel-efficient vehicles carrying greater payloads, pushing the boundaries of exploration [8]. This immediately translates to cost reductions, allowing us to reach further and achieve more with each mission. For instance, a 10% reduction in spacecraft weight could translate to millions of dollars saved in launch costs. Additionally, graphene's thermal conductivity excels at regulating internal temperatures, a critical factor for astronaut comfort while protecting sensitive electronics during long journeys through harsh environments (refer to Table 1) [9].

Beyond structure, graphene's unique electronic properties suggest superior radiation shielding [10]. This is vital for safeguarding astronauts venturing into deep space, where exposure to harmful radiation is a significant threat. Finally, graphene's light absorption and conductivity make it suitable for developing highly efficient solar cells, providing a reliable power source for spacecraft venturing beyond the reach of sunlight.

Ongoing research is exploring novel ways to scale graphene production efficiently, which could accelerate its adoption in commercial applications. Additionally, graphene can be susceptible to degradation in harsh space environments. Further research is needed to ensure its long-term stability. Finally, while graphene offers significant benefits, its current production cost is high. Overcoming these challenges will require sleuthing by researchers, meticulously examining every aspect of the production process to identify areas for improvement. Developing cost-effective manufacturing methods and integrating graphene effectively into existing spacecraft systems are crucial for practical implementation.

Despite these challenges, graphene holds colossal promise for revolutionizing space exploration and tourism. Its unique properties pave the way for lighter, more efficient spacecraft, improved thermal management, superior radiation protection, and reliable power generation. Continued research and development are key to unlocking graphene's full potential in this exciting domain.

Alternate Energy: A Bridge to Sustainability

With Graphene's potential for clean energy applications, how can we encourage collaboration between researchers, governments, and private industries to accelerate its development? We have already shown that graphene can pave the way for a more sustainable energy future. To fully realize this potential, we must foster an ecosystem of innovation that transcends traditional boundaries. This calls for a multi-pronged approach: establishing international research consortia, creating incentive structures for private sector investment, and implementing supportive policy frameworks. By aligning these efforts, we can catalyze breakthroughs in graphene-based energy solutions, potentially revolutionizing everything from solar cells to energy storage systems. The time to act is now, our planet and humanity's future depends on our ability to harness such groundbreaking technologies.

Moreover, its renowned properties, such as high surface area and conductivity make it very suitable for developing high-performance supercapacitors and batteries. Wouldn't it be remarkable if efficient energy storage for renewable sources, like solar and wind power, could readily enable a more reliable and sustainable energy grid while leaving a carbon-negative footprint?

Studies suggest graphene-based supercapacitors could store ten times more energy than their traditional counterparts [11][23]. The need for alternative energy solutions like tidal and geothermal power is more pressing than ever. Underground robotics, powered by advancements in AI and material science, can play a crucial role in developing and maintaining these sustainable energy sources.

For instance, graphene-based solar cells have the potential to be more efficient than traditional silicon counterparts, leading to a significant increase in solar energy generation and a reduction in our dependence on fossil fuels [12]. Consider solar panels generating 20% more electricity; that's a substantial leap towards a cleaner energy future. Because our oceans face an existential threat from

---

[23] Graphene-based supercapacitors are an emerging technology with promising potential for significantly improving energy storage capabilities.

pollution, graphene's exceptional filtration properties offer a glimmer of hope. Specialized filters utilizing graphene could efficiently remove pollutants and excess CO_2 from our oceans, contributing to a cleaner and healthier marine environment.

Beyond generation and storage, graphene can act as a catalyst for clean energy technologies. Its ability to efficiently facilitate chemical reactions makes it a promising candidate for hydrogen production, a clean and sustainable alternative to fossil fuels. By developing efficient electrocatalysts based on Graphene, we can pave the way for a future powered by clean and renewable hydrogen fuel. Finally, Graphene membranes show promise for developing more efficient and energy-saving desalination processes [13]. AI can be a powerful tool for environmental remediation. By analyzing vast datasets and simulating various scenarios, AI can help us develop and optimize strategies for pollution cleanup. This includes tackling historical pollution from the Industrial Revolution, a monumental challenge requiring innovative solutions.

By embracing continuous improvement (KAIZEN), we can leverage AI to address these environmental challenges effectively. Access to clean drinking water is a growing concern globally. By creating more efficient desalination methods, graphene can contribute to a sustainable future by providing clean water in regions facing water scarcity.

Computing Architecture: New Design Paradigm

Traditional CPUs have been the workhorses in the digital world, though they've truly struggled with the massive parallel computations required for tasks like image recognition or natural language processing [14]. This is where a recently emerging paradigm comes in.

Neural Processing Units (NPUs) are specialized hardware designed specifically for the unique demands of AI workloads and are on the rise. Built for high-performance, low-power execution of massive parallel calculations, NPUs are significantly faster and more energy-efficient for AI applications compared to CPUs.

For decades, Moore's Law, which predicted the exponential growth of transistors on microchips, fueled advancements in processing power. However, we're nearing its physical limits. NPUs offer a new path forward, achieving significant performance gains without necessarily adhering to traditional transistor density scaling. While GPUs have demonstrably tackled some AI tasks with their parallel processing capabilities, they weren't designed specifically for the unique mathematical operations used in deep learning algorithms. NPUs, on the other hand, are built from the ground up for AI, delivering superior performance and efficiency compared to even the most powerful GPUs.

This is also where graphene enters the exciting world of computing architecture. Its exceptional electrical properties hold immense promise for revolutionizing NPU design. Graphene's superior electron mobility could enable significantly faster computations within NPUs, leading to quicker training times for AI models and faster execution of complex tasks [15]. Additionally, its low electrical resistance could lead to more energy-efficient NPUs, a crucial factor for large-scale AI deployments. The potential for smaller transistor sizes with graphene paves the way for even more compact and powerful NPUs. Imagine highly efficient AI-powered devices at the network's edge, closer to where data is generated. But graphene's impact extends beyond NPUs.

Its unique properties can improve various aspects of digital infrastructure, from faster data transfer within computer systems to flexible and wearable electronics seamlessly integrated into everyday wear[24]. The future of computing lies in collaboration; specialized hardware like NPUs working together with advanced materials like graphene to unlock the full potential of AI. This collaborative approach, driven by continuous research and innovation, holds the key to a future powered by efficient, intelligent computing.

Challenges exist, sure. Large-scale production of high-quality graphene remains an ongoing challenge for widespread use in energy technologies [16]. Similarly, cost reduction is essential for

---

[24] Graphene's flexibility enables self-correcting chips to maintain optimal performance in fluctuating conditions.

making graphene-based solutions commercially viable. Additionally, effectively integrating graphene into existing energy infrastructure requires careful planning and research to maximize its impact. Finally, the environmental impact of graphene production needs further assessment to ensure a truly sustainable solution.

Graphene offers exciting possibilities for bridging the gap towards a sustainable energy future [17]. Its unique properties can lead to more efficient energy storage, generation, and clean water production. However, addressing scalability, cost, and environmental considerations is key to its successful implementation.

Chapter 7

A Giant Leap For Mankind

Blood-red sunlight spilled across Tokyo's skyline. Elena's coffee had gone cold, forgotten during the endless night of preparation. Her hands shook slightly as she straightened papers that didn't need straightening.

Today would change everything.

The command center buzzed with whispered conversations and the soft clicking of keys. Anya hadn't slept in thirty hours, her dark curls wild around her face as she muttered to her screens. Kai paced between power stations, touching each one like a worried parent checking on fever-stricken children.

Jake sat in their midst now, no longer a digital ghost but a man with tired eyes and a coffee addiction worse than Elena's. He'd brought donuts this morning. Nobody had eaten them.

"Got him."

Reeves' voice cut through the quiet like a gunshot. They all turned. He stood in the doorway, jacket rumpled, looking like he'd aged ten years overnight.
"Vladivostok," he said. "Old research lab. Local police are ready."
Elena looked at her team; really looked at them. Anya's chewed-up pencil, Kai's lucky shirt with the coffee stain, and Jake's fingers drumming an endless rhythm on his desk. These weren't just colleagues anymore. They'd become something else during these endless months of battle.

"Remember when we thought this was just about better batteries?" she asked softly.
Kai actually laughed. "God, we were young."
They moved like they'd rehearsed it a thousand times. Reeves coordinated with strike teams in four languages while their Bridge came alive, spreading its net across the digital world. Jake's quantum encryption wrapped around them like armor.
Raj's voice crackled through from Bangalore, warm despite the static: "Drones are hot. These babies will light the way through hell itself."
Anya hunched over her keyboard, muttering as she tracked Novak's digital footprints. "Come on, you arrogant bastard. Show me where you're hiding."
"Power reserves at twelve hours," Kai reported, his usual jokes nowhere to be found. "After that..." He shrugged.

The assault began in Vladivostok's pre-dawn gray. Their teams moved like shadows, graphene tech making them nearly invisible. Novak's men fought back hard; they were still human, still someone's sons and brothers, even if they fought for the wrong side.
"The generators!" Jake's coffee cup crashed to the floor as he leaped up. "Take out the auxiliary power. It's his weak spot!"
The order went out. Seconds stretched like years.

Novak's face filled their screens, his eyes burning with a zealot's fire. "You really think you've won? "His laugh sounded like breaking glass. "Systems like yours always fall because they're too idealistic, too fragile."

Elena stood. Her voice stayed steady even as her heart hammered. "You never understood, did you? Dreams don't shatter. They evolve.

You lost because you stood alone. We won because we stood together."
His reply was cut short by the sound of doors breaking. Shouts in Russian. The screen went black.

"Target secured." Reeves' voice was hoarse. "Novak is in custody."
The cheer that went up shook dust from the ceiling. Anya hugged Jake, then looked mortified. Kai collapsed into his chair, laughing and crying at once.
But Elena raised her hand. "This isn't the end," she said. "It's barely the beginning. We haven't just built a bridge. We've built a promise."
The next days blurred together. Jake worked openly with them now, his mysterious past becoming funny stories over late-night takeout. They hunted down Novak's digital traps one by one, each victory small but vital.

When Raj arrived from India, he brought samosas and a smile that lit up the room. He found Elena on the roof, watching the sunset.
"You did it," he said simply.
Elena watched her team through the window below. Anya was teaching Jake some algorithms while Kai tried to balance a pencil on his nose. Reeves pretended not to smile.
"No," she said. "We did it. And we're not done yet."

The Tokyo sky blazed with color like the universe itself was celebrating. Somewhere below, their Bridge hummed, ready to connect a world of strangers into a family of dreamers.

They had won the war with blood and sweat and coffee and terrible jokes. Now came the hard part.

Now they had to build tomorrow.

...

Looking Ahead

As groundbreaking technologies reshape the world, the question arises: who will guide us toward a brighter future? Leadership in this era transcends brilliant scientists and engineers like Anya, Dr.

Petrova, and Kai. It extends to policymakers, educators, and every citizen who can influence the responsible development and application of these tools. Ensuring equitable access and minimizing the digital divide is no longer optional, it is an ethical imperative. Decisions made today will determine whether these innovations uplift humanity or exacerbate inequalities[25].

Yet, these advances come with significant challenges. Quantum computing faces immense scalability challenges, AI raises critical concerns about bias and misuse, and graphene production still struggles with cost and environmental impacts. Policymakers and innovators must address these barriers collaboratively, ensuring the technologies' benefits outweigh their risks. With vigilance and intentional action, we can shape a future that is inclusive, sustainable, and impactful.

The Future is Now

The once-distant future of quantum computing, AI, and graphene is here, disrupting industries and challenging conventional thinking. Far from the realm of science fiction, these technologies are unlocking possibilities that once seemed unattainable. Graphene's unparalleled conductivity and strength, paired with quantum computing's ability to solve problems of unimaginable complexity, represent a renaissance of human innovation.

Meanwhile, AI enables systems to adapt and learn, pushing boundaries in automation and decision-making. With these advancements comes immense responsibility. Its transformative potential is tempered by concerns over ethical use, data privacy, and societal impacts. Quantum computing faces technical and practical obstacles before becoming a widely accessible tool. Similarly, graphene's industrial scalability remains a challenge, particularly in ensuring environmentally sustainable production. Addressing these issues head-on is not just an option but a necessity. By navigating

---

[25] Leadership in the quantum era extends beyond technology creators. Policymakers, educators, and citizens also play vital roles in guiding the ethical use and equitable distribution of innovations.

these challenges responsibly, leaders can unlock the transformative potential of these technologies for the benefit of all.

The 361 Degree Leap Forward: The Quantum Mindset

The classical worldview: linear, siloed, and predictable, has served humanity well, but its limits are clear in the face of quantum possibilities. True brilliance lies in the leap into quantum thinking, where superposition, entanglement, and parallelism defy traditional logic and open new frontiers.

To unlock these opportunities, leaders and innovators must transform their mental models. Embracing ambiguity and exploring multiple possibilities simultaneously is key. Decentralized teams and iterative experimentation offer advantages, enabling rapid adaptation to change. However, these approaches also introduce complexities in coordination and accountability. Effective leadership will require balancing innovation with practicality and fostering environments that encourage curiosity and diverse perspectives.

The organizations of the future will resemble adaptive ecosystems rather than rigid hierarchies. Decentralized networks, agile methodologies, and empowered teams will define success in an era of constant change. Transforming our thinking is not just philosophical; it is a necessity to navigate the complexities of the quantum age. Leaders who embrace this mindset will drive the most meaningful breakthroughs, redefining what is possible while ensuring these advancements serve humanity.

Embracing the full potential of deep innovation requires courage, creativity, and a willingness to challenge conventional assumptions. Ambiguity should be celebrated, not feared, as it often sparks the most transformative ideas. The profound journey toward graphene and quantum-inspired solutions demands leaders who empower teams to explore uncharted territories, encouraging experimentation and valuing diverse perspectives.

Yet this journey is fraught with obstacles. Transformative technologies can disrupt industries and displace workers, creating economic and social challenges that must be addressed.

Decentralized networks, while powerful, may also introduce vulnerabilities that demand careful oversight. As Arthur C. Clarke famously said, "Any sufficiently advanced technology is indistinguishable from magic." Yet the true magic lies in the willingness to think differently, rewire assumptions, and lead boldly into the quantum revolution. By confronting both the possibilities and the challenges, innovators will not only redefine what is possible but also ensure their advancements serve humanity. This is not just the next step, it is a giant leap forward and it has only just begun.

References

1. Ahluwalia, P., & Mittal, N. (2021). Internet of Things Concept and Its Applications. In D. Gupta, H. C. de Albuquerque, V., A. Khanna, & P. L. Mehta (Eds.), Smart Sensors for Industrial Internet of Things (pp. 7-36). Springer.

2. Akter, H., Jahan, S., Saha, S., Faisal, R.H., & Islam, S. (2022). Evaluating Performances of VPN Tunneling Protocols Based on Application Service Requirements. In M.S. Kaiser, K. Ray, A. Bandyopadhyay, K. Jacob, & K.S. Long (Eds.), Proceedings of the Third International Conference on Trends in Computational and Cognitive Engineering.

3. Brooks, F. P. (1987). "No silver bullet: Essence and accidents of software engineering." In Computer (Vol. 20, No. 4, pp. 10-19). IEEE.

4. Brundage, M., Mitchell, S., Prabhu, M., Scheuerman, B., & Wagstaff, G. (2018). The malice of intent in artificial intelligence. Science, 361(6406), 806-810.

5. Caruana, J. M., et al. (2019). "Graphene-based photonic integrated circuits for quantum communication applications." Nanophotonics, 8(15), 6303-6316.

6. Chen, J., Liu, B., & Gao, X. (2020). Thermal properties of graphene-based polymer composite materials: A molecular dynamics study. Results in Physics, 16, 102974.

7. Diamandis, P. H., & Kotler, S. (2012). Abundance: The future is better than you think.

8. Dodda, A., Subbulakshmi Radhakrishnan, S., Schranghamer, T.F. et al. Graphene-based physically unclonable functions that are reconfigurable and resilient to machine learning attacks. Nat Electron 4, 364-374 (2021).

9. Gerstner, E. (2010). Nobel Prize 2010: Andre Geim & Konstantin Novoselov. Nature Physics, 6, 836.

10. Geim, A. K., & Novoselov, K. S. (2007). "The rise of graphene." Nature Materials, vol. 6, no. 3, pp. 183-191.

11. Graimann, B., Allison, B. Z., & Pfurtscheller, G. (2010). Brain-Computer Interfaces: Revolutionizing Human-Computer Interaction.

12. Guichard, Ami. Automobile Year 1958 Edition. EDITA S.A., 1958.

13. Heikkilä, M. (2023). Making an image with generative AI uses as much energy as charging your phone. MIT Technology Review. Available at: https://www.technologyreview.com/2023/12/01/1084189/making-an-image-with-generative-ai-uses-as-much-energy-as-charging-your-phone

14. Ikram, M., Maqsood, A., & Bashir, A. (Eds.). (2023). Graphene: A wonder material for scientists and engineers.

15. Kao, S.-C., Parashar, A., Tsai, P.-A., & Krishna, T. (2022). Demystifying Map Space Exploration for NPUs. In [arXiv preprint arXiv:2210.03731] (pp. 1-18).

16. Karpathy, A. (2017). A Primer on Deep Learning Models and How They Work.

17. Kramp, T., Kranenburg, R., & Lange, S. (2013). Introduction to the Internet of Things. In Enabling Things to Talk (pp. 1-10). Springer.

18. Lu, Y., Liu, X., & Kuzum, D. (2018). Graphene-based neurotechnologies for advanced neural interfaces. Current Opinion in Biomedical Engineering, 6 (1), 138-147.

19. Miller, T., et al. (2019). "The Importance of Explanations in Artificial Intelligence and Machine Learning." Proceedings of the National Academy of Sciences, vol. 116, no. 13, pp. 6231-6236.

20. Mirabedini, A., Ang, A., Nikzad, M., Fox, B., Lau, K.-T., & Hameed, N. (2020). Evolving strategies for producing multiscale graphene-enhanced fiber-reinforced polymer composites for smart structural applications. Advanced Science, 7(8), 1903501.

21. Novoselov, K. S., Geim, A. K., Morozov, S. V., Jiang, D., Zhang, Y., Dubonosov, S. V., & Firsov, A. A. (2004). Electric field effect in atomically thin carbon films. Science, 304(5673), 102-104.

22. Pang, C., Sun, J., Fan, Z., & Xie, Z. (2017). Highly sensitive and flexible pressure sensors based on mesoporous silica nanoparticles decorated graphene oxide. ACS Nano, 11(12), 11768-11777.

23. Santos-Ceballos, J.C., Salehnia, F., Romero, A. et al. Application of digital twins for simulation based tailoring of laser induced graphene. Sci Rep 14, 10363 (2024).

24. Schedin, F., Geim, A. K., Morozov, S. V., Hill, E. W., Blake, P., Katsnelson, M. I., & Novoselov, K. S. (2007). Detection of individual gas molecules adsorbed on graphene. Nature Materials, 6 (9), 652-655.

25. S. Sutar, A. Raha and V. Raghunathan, "Memory-Based Combination PUFs for Device Authentication in Embedded Systems," in IEEE Transactions on Multi-Scale Computing Systems, vol. 4, no. 4, pp. 793-810, 1 Oct.-Dec. 2018.

26. Tarasov, K. (2024). How the massive power draw of generative AI is overtaxing our grid. CNBC. Available at: https://www.cnbc.com/2024/07/28/how-the-massive-power-draw-of-generative-ai-is-overtaxing-our-grid.html

27. Thompson, K. (1984). "Reflections on trusting trust." In Communications of the ACM (Vol. 27, No. 8, pp. 761-763). ACM.

28. Vaswani, A., Shazeer, N., Parmar, N., Uszkoreit, J., Jones, L., Gomez, A. N., Kaiser, E., & Polosukhin, I. (2017). Attention is all you need.

29. Wang, Y., Wang, C., Gu, C., Cui, Y., O'Neill, M., & Liu, W. (2019). Theoretical Analysis of Delay-based PUFs and Design Strategies for Improvement. ISCAS 2019 - Proceedings Article 8702722 Institute of Electrical and Electronics Engineers Inc.

30. Wassei, J. K., & Kaner, R. B. (2010). "Graphene, a Promising Transparent Conductor." Materials Today, vol. 13, no. 3, pp. 52-59.

31. Xin, N., Lourembam, J., Kumaravadivel, P. et al. (2023). Giant magnetoresistance of Dirac plasma in high-mobility graphene. Nature, 616, 270-274.

32. Yehoshuva, C., Raja Adhithan, R., Nalla Anandakumar, N. (2021). A Survey of Security Attacks on Silicon Based Weak PUF Architectures. In: Thampi, S.M., Wang, G., Rawat, D.B., Ko, R., Fan, CI. (eds). Communications in Computer and Information Science, vol 1364. Springer, Singapore.

33. A Graphene-Based Straintronic Physically Unclonable Function. Subir Ghosh, Yikai Zheng, Shiva Subbulakshmi Radhakrishnan, Thomas F Schranghamer, and Saptarshi Das. Nano Letters 2023 23 (11), 5171-5179.

34. European Commission. (2024). Regulation (EU) 2024/1689 on artificial intelligence. Available at: https://eur-lex.europa.eu/legal-content/EN/TXT/?uri=CELEX:32024R1689

35. IBM Security. (2023). Cost of a Data Breach Report 2023. Available at: https://www.ibm.com/reports/data-breach

36. Identity Theft Resource Center. (2023). 2023 Data Breach Report. Available at: https://www.idtheftcenter.org/publication/2023-data-breach-report

Afterword

When Kireet Kokala invited me to write the afterword for *The Graphene Bridge*, I initially hesitated—not due to doubts about the book's appeal, but because I wasn't a subject matter expert. Although I work with and respect professionals who possess expertise in these areas, particularly engineers who conduct research in machine learning and artificial intelligence, I have never read a book that cuts to the heart of defining what these often intimidating topics are, how they are integrated in a tapestry of knowledge to benefit society, and why a layperson should care about graphene or the interdisciplinary nature of the topics presented in this book. For that reason, I approached reading cautiously, eager to offer a new perspective that would engage readers outside the specialized fields of AI, quantum computing, and atomic batteries.

What stood out distinctly was the authors' unique integration of a global narrative that infused emotional depth into the realities of advancing technology, carefully addressing the human, fiscal, and intellectual implications. As a novelist and engineering educator, I appreciated the creativity to transform knowledge in ways that professors Chris Golde and Timothy Dore refer to as transformation, or teaching broadly and possibly outside traditional classroom environments. How would one communicate how a material like graphene fundamentally alters entire industries, or perhaps even create entirely new markets?

This book is timely since graphene presents immense opportunities for innovation and abuse. As the United States and global economies change, almost every part of society will be touched, including healthcare (e.g., faster diagnostics and personalized care); cybersecurity; energy storage, and global communications and data protection.

Without proper oversight and regulation of the work presented in this book, however, high costs, bias, and misuse of technologies could place people in harm's way. The authors warn us to take heed to technological hurdles for scaling the production of graphene and communicating the value of graphene across industries and to consumers. Like any innovation, policies must be created and evaluated to address the complexity of sustainability, data privacy, and ongoing stewardship of resources impacted by advancements with graphene.

As AI, quantum computing, and atomic battery technologies rapidly evolve, and the ethical considerations of the concepts presented in this book deepen, digestible books like this are increasingly necessary. They will enable inquisitive learners to grasp what I frequently challenge my students to uncover—the critical "So What?" of any given topic.

The Graphene Bridge is unique, combining fiction and nonfiction to bridge complex concepts that might otherwise be daunting if studied independently or outside a course. As a professor accustomed to reading intricate information and prompting students to distill complex topics into clear, concise ideas, I greatly appreciated how the authors introduced and defined technical terms and provided diverse applications and integrations of those terms for a broader audience.

Ultimately, the authors' thoughtful exploration inspires a compelling vision for our collective future. As graphene reshapes industries and opens novel pathways, each of us—practitioners, technologists, investors, consumers, and leaders alike—must thoughtfully consider our roles in responsibly steering these advancements toward the betterment of society.

By Dr. Monica Cox
Distinguished Professor of Engineering, The Ohio State University, and 9 Time Author

Author - Kireet Kokala

As the visionary behind The Graphene Bridge, Kireet Kokala's decade-long fascination with graphene's immense potential drives the book. His research and passion for the material have been pivotal in this endeavor. Leveraging his comprehensive background and significant roles at NetApp and Hewlett-Packard, including his public speaking expertise, Kireet ensures that complex concepts are presented clearly and engagingly for readers from all walks of life. In his present role as Director of Hyperscaler Data Science at Hitachi Vantara, Kireet leads technological innovation and is instrumental in pioneering data science initiatives that advance R&D and artificial intelligence applications. Kireet's leadership is marked by a commitment to innovation, inspiring his peers to push the boundaries of what is possible.

Largely growing up in Atlanta, Georgia, Kireet's journey is deeply rooted in its vibrant cultural environs. He enjoys hiking, DIY projects, and exploring new places. Kireet is a multilingual poet and an influential public speaker, honing his communication and leadership skills as a Distinguished Toastmaster with Toastmasters International. He has broad expertise in technology and biological sciences, holding an MS in Computer Science from Georgia State University with a focus on Natural Language Processing, and a BS in Biology (pre-medicine) from The University of Georgia.

For more information on the book or to contact me, visit: kireetverse.com

Kireet's writings on technology and leadership are featured in industry publications, establishing him as a pioneering innovator. In his free time, Kireet connects things full circle by engaging with his community, exploring nature, spending quality time with his family, and enjoying a harmonious balance between his personal and professional passions.

Author - Supantha Banerjee

Supantha Banerjee is a seasoned global operations and technology leader with a passion for digital innovation. He has held executive positions at prominent companies including Technology Partners, EyeCare Partners, Aspen Dental, and Macy's. His expertise lies in transforming digital strategies into practical solutions that enhance consumer experiences and drive business growth.

A thought leader in technology and innovation, Supantha's thought leadership and articles have been featured in CIO.com and other leading industry publications. He has contributed to numerous white papers and research projects, sharing his insights on innovation and operational excellence.

Supantha holds a Bachelor of Engineering degree from the National Institute of Technology, Durgapur, India, and an MBA from the University of Michigan, Ann Arbor. When not shaping the future of technology, he can be found mentoring aspiring leaders, playing music, or cheering for his favorite cricket and soccer teams. Through his work and pursuits, Supantha continues to harmonize the realms of technology, leadership, and creativity, inspiring others to embrace innovation in both their professional and personal lives.

www.ingramcontent.com/pod-product-compliance
Lightning Source LLC
LaVergne TN
LVHW022110200926
848481LV00003B/1131
9798998548307